ZOOLOGY

Introduction to
ZOOLOGY

by

THEODORE H. SAVORY
M.A., F.Z.S.

Formerly Exhibitioner of
St. John's College, Cambridge

Illustrated by
MELCHIOR SPOCZYNSKI

PHILOSOPHICAL LIBRARY INC.

15 EAST FORTIETH STREET

NEW YORK 16, N.Y.

Published 1968 by Philosophical Library Inc.,
15 East 40th Street, New York 16, N.Y.

Printed in Great Britain for Philosophical Library
by Butler & Tanner Ltd, Frome and London

CONTENTS

PREFACE

The writer of any introductory science book is faced at the outset with two problems.

The first of these is to decide, as reasonably as possible, what is to be omitted. A complete, comprehensive Zoology might be expected to fill thirty thick volumes, between which and the size of this book there is a considerable difference. I have chosen to neglect the microscopical sciences of histology and embryology, the physico-chemical science of physiology, and the statistical science of genetics because these branches of biology deserve separate volumes in this series. And I have omitted phenomena that may be said to belong to animal biology rather than to pure zoology, so that there is little discussion here of such topics as symbiosis, parthenogenesis, behaviour and parasitism.

For there comes, and this is the second difficulty, a desire to expound the scientific structure of an observational study such as zoology. The basic function of every science is the arrangement of its facts and the ordering of its ideas in a logical system. I have therefore stressed the systematic and evolutionary sides of zoology, and have used the principles of taxonomy as a helpful guide through the many mysteries with which animals intrigue us, frustrate us and fascinate us.

In so doing I have included descriptions, albeit sometimes rather short, of many orders and classes of animals which most students of elementary zoology customarily neglect. My mention of such groups as Pycnogonida and Pogonophora and others emphasizes the almost incredible diversity of animal life, and must help to form a truer picture of the world of animals.

Animals are organisms which, like ourselves, have to protect and feed themselves and reproduce their kind, living wherever they can and however they can, in a bewildering variety of ways, and achieving their ends by a remarkable number of adaptive means; but in all this variety there lies a fundamental uniformity which the systematist detects and expresses as a coherent science.

This is indeed the essence of scientific zoology as I see it, and which I have tried to make acceptable to those, rapidly increasing in number, who feel a compelling interest in animals all over the world. Perhaps the greatest problem of all is satisfactorily to join this inquisitive interest to the serious processes of systematic science.

T.H.S.

Kensington

PART ONE

INTRODUCTORY ZOOLOGY

CHAPTER ONE

APPROACH TO ZOOLOGY

Zoology, the scientific study of animals, offers to the reader a subject of surpassing interest, to the student a number of problems of the widest diversity, and to the writer a topic of infinite fascination. Zoology owes these qualities to the animals themselves, their numbers and their variety, their strange forms and their stranger habits, their unquestioned value to ourselves and their equally undoubted hostility as pests and parasites.

Men have studied animals from earliest times and over the years have developed a tripartite science. There is the zoology of the laboratory, where structure and function are investigated, there is the zoology of the field, where there is revealed the behaviour of the living animal, and there is the zoology of the study, where the puzzled zoologist tries to systematize his observations and to draw significant conclusions from his discoveries.

The foundation of all zoology has always been morphology, the science of structure. Until little more than half a century ago it was almost the only branch of zoology to receive attention. Other sub-sciences were but slightly developed; they provided footnotes or postscripts to certain aspects of morphology, or they were the concern of a few specialists, but they were not prominent enough to stand out as being worthy of the attention of all. Gradually, the voices of the geneticist, the ecologist and others made themselves heard in criticism of the over-stressed tyranny of laboratory work. Biology, it used to be said, is the science of life, but its followers seldom see an animal alive; their first action is to kill it and then to content themselves with an examination of its corpse. This, it was ironically added, is necrology, the science of the dead.

There was some justification for this criticism, which is still to be heard on occasion, and as the younger sciences grew the pendulum began to swing to the opposite side. Morphology was even decried and threatened with a subordinate position and a smaller fraction of zoological esteem. But in nearly sixty years this has not come to pass and there is still the need for an accurate grounding in comparative anatomy for all who aspire to the status of competence in zoology. The present position is simply this: not that we should regard morphology less, but that we should regard its offspring more.

As this has been done, zoologists, following their different inclinations in their different ways, have developed an observational, descriptive science which is clearly contrasted with the experimental, quantitative sciences of physics and chemistry. Zoology, in other words, is a science that is not based on measurement and calculation but, instead, on observations and speculation. This is, indeed, its fundamental characteristic.

It is a characteristic that is shared by other sciences, as old, as well-established as zoology. Botany is its most obvious parallel; geology, and in particular physiography, are dependent wholly on observation; so too are paleontology and archaeology. When a geologist repairs to his laboratory, he occupies himself with physical or chemical observations; he can no more carry out experiments with plateaux or flood plains than an astronomer can 'correct old Time and regulate the Sun'. The paleontologist is compelled to dig, and there is no real difference between the labour of a man who digs for fossils and one who digs for buried cities and shards of pottery. The latter is an archaeologist and would not claim to be an experimental scientist, and his thoughts and those of his zoological colleagues are likely to turn to the museum rather than to the laboratory.

If, therefore, the observational character of zoology and of several other sciences may be taken as established, the conclusion should be followed by a correction of the commonly held idea that the quantitative sciences are in some way superior to the others.

There can be no inherent superiority of any branch of knowledge over any other, though it is true that some knowledge makes a very inconspicuous contribution to the lives of men, and some makes contributions that are spectacular. But there is no scale which measures the relative values of different kinds of knowledge, which are distinct only in the occasions and opportunities which bring them to our notice.

It follows that in zoology reasoned speculations take the place of the formal hypotheses of the physical sciences. In the familiar 'scientific method' a hypothesis is used to suggest unsuspected facts which can be proved by experiment. Zoological speculations can seldom be tested experimentally; they can be brought to trial only by comparison with other speculations. Morphology, for example, mentioned above as the foundation of all zoological knowledge, is not the science of anatomy but the science of comparative anatomy. In other words, much that is best in zoology is more closely related to philosophy than to technical research.

Zoology and botany are the chief components of the inclusive subject which Lamarck in 1802 named biology, the science of all living things. Certain characteristics distinguish animals from plants, and the existence of limited exceptions only emphasizes the variety of living things, whatever they may be and wherever they may live.

The fundamental difference between animals and plants is based on the energy relationships between the organism and its surroundings or environment. Machines are devices for the conversion of energy from one form to another: animals are not machines, but they may often be studied as if they were, because they show a similar capacity for energy conversion.

Plants can absorb the radiant energy of the sunshine with the help of the chlorophylls and other pigments that their cells contain. This enables them to build up, or synthesize, large endothermic, or energy-containing molecules from the smaller molecules of compounds like carbon dioxide, water and potassium nitrate. These compounds when oxidized liberate energy which the plant uses to produce more tissues, and so to increase its size and to produce seeds or spores, and thereby to reproduce its kind. Plants make very slow movements of their parts and move from place to place hardly at all; in consequence they accumulate more energy than they consume and become stores of potential energy, available for other purposes or other organisms.

Animals possess only a very slight power of making use of solar radiation. They take in supplies of energy in the form of food, liberate the potential energy from this, usually by a direct oxidation process under the control of enzymes, and make use of this energy in growth, reproduction and movement. Much of the energy appears finally in the form of heat, when it is lost to the environment. Thus animals dissipate energy in contrast to plants, which store it; whence it follows that all animal life is now dependent on plant life. The familiar phrase, 'All flesh is grass', was not used by Isaiah with this implication in mind, but it is often quoted by biologists as an apt, concise expression of the basic nature of animal metabolism.

In addition to this, there are other characteristics of animal life which a zoologist bears constantly in mind. Animals, perhaps more conspicuously than plants, are very closely related to their environment, and in any comprehensive view cannot be considered apart from it. The environment is in a constant state of change, these changes impinge on the animal and the animal reacts or responds to them. These reactions are remarkable because of the way in which they tend towards the survival of the individual and of the race; they are selfish responses, of advantage to the animal itself. It is only because of this that survival is possible. One of the surprises awaiting the zoologist who, until he has had it pointed out to him, has seen a world teeming with an apparently flourishing multitude of animals, is the fact that the animal is an amazingly vulnerable object. Its life is so easily lost that it survives only by virtue of this selfish quality in its responses to an environment that constantly threatens it with extinction.

Thus the responses of an animal are responses to changes in its environment, the adaptations of animals are adaptations to their particular environment; the relation between animal and its environment is what animal life is. The two touch at a thousand points; they are inseparable and together they form a related whole, the organic unit. The laboratory zoologist who deals too exclusively with chloroformed frogs, formalined dogfish and embalmed rabbits, is in danger of forgetting this. He is as one who has contented himself with the study of a concave mirror and is confident that he knows everything about its powers of image formation, yet all the time his mirror has its convex side, showing quite different powers, and the two are inseparable.

All the ideas, the concepts, that are suggested by paleontology, embryology, biochemistry, comparative anatomy, ethology and ecology must be fitted into their appropriate places in the fundamental theory of all living things, the theory of organic evolution. The detection and description of the evolutionary relationships between the various groups of animals, a study that is known as phylogeny, produce most of the problems that puzzle or fascinate zoologists today. Some of these questions will be found in the pages that follow, where they intentionally receive at least as much prominence as the facts of morphology.

One of the fundamental purposes of all sciences, whether physical or biological, is to arrange facts and ideas in a system. This is systematics, the most scientific of all the sciences.

Scientists of every kind are agreed that their investigations have revealed an orderliness in the works of nature and it is significant that this is true of every branch of science, physico-quantitative or biologico-descriptive, and that progress has always become more rapid as soon as this orderliness has been established and the system has been recognized. Science can deal only with phenomena so ordered that they become predictable.

There have been thoughtful writers who have asked whether this universal orderliness of nature is a real, an objective feature of the world, or whether it is a subjective feature, abstracted from the total of universal phenomena and existing only in the minds of the scientists. Whatever the answer, the consequences seem to be immaterial, for it is only in the minds of the scientists that it exists at all, and only by the thoughts of scientists that scientific progress is made. The orderliness may be apparent in no more than a series of pointer-readings on an arbitrary scale, and yet scientific discussion of the phenomena thus recorded at once becomes possible.

Again, whatever the answer, the orderliness of nature is universal. Systematized phenomena everywhere, physical, chemical or biological, are so unquestionably the material of science that the descriptions 'scientific' and 'systematic' become interchangeable or synonymous.

The nature of the science of zoology and its position among the other sciences should now have become apparent. Whereas the data of physical science can be readily summarized in laws that lead to hypothesis, the data of zoology tend rather to suggest helpful notions, fertile concepts, or, more simply, good ideas. In other words, a zoologist works with the help of concepts where a physicist works with hypotheses; and the most significant difference, if there is any difference at all, is that the good hypothesis suggests fresh experiments, while a fertile concept illuminates other observations, both old and new.

This introductory chapter may therefore be summarized by setting forth the purposes of systematics in zoology in six paragraphs:

1. To arrange all animal species in their appropriate places in a plan or scheme which will make plain their various resemblances, differences and relationships.

2. To construct such a scheme to demonstrate the probable course of evolution in the animal kingdom, and so to illustrate the nature and origins of the differences and resemblances.

7

3. To distinguish such a scheme from any artificial classification such as carnivora, herbivora, omnivora; aquatic, terrestrial, aerial; birds, beasts, fishes; as of no scientific significance.

4. To give a name to every species so that it may be quickly and certainly recognized by readers; and to give names to all genera and other groups to which species may be assigned.

5. To present such a scheme so that any specimen can be easily and quickly placed in its correct position, and its name, if known, be accurately determined.

6. To expound the above items so that they may be readily grasped by students, readers and others; and generally accepted by the leaders of zoological thought.

All this amounts to a considerable body of theoretical zoology, deduced from the facts observed in the laboratory and the field, and it accurately summarizes the nature of zoology as a branch of natural science.

PART TWO
SYSTEMATIC ZOOLOGY

CHAPTER TWO

THE PRINCIPLES OF CLASSIFICATION

There are very many different animals. This remark, with which every zoologist must agree, could be expressed by saying that the chief features of animal life are its multiplicity and its diversity. These are among the first obstacles to the study of zoology, and they are attacked in different ways.

First, as to multiplicity. The described species number more than a million, a number which is almost meaningless to the ordinary mind, even though population problems and national finances have made us all increasingly familiar with the word. Multiplicity is usually met by specialization: a man may usefully spend a lifetime on a group of a thousand or only a few hundred species.

The diversity is more conspicuous and more intractable. Even within the limits of a single class, such as the mammals, there are contrasts such as the whale and the shrew, and over the whole range of the animal kingdom the differences in size, structure, habits and behaviour between elephants and eel-worms, bats and barnacles, ospreys and oysters are so vast that one is amazed at the power of the life principle to invest a range of forms almost beyond the imagination of man.

Complexities such as these can be countered only by a system of classifying, so that zoological classification appears partly as a human necessity, a consequence of the limitations of the intellect. Classifying is also a fundamental characteristic of us all, so that there is no surprise that from the time of Aristotle onwards naturalists have shown a tendency to classify as an almost instinctive reaction to the complexity of their studies. Through the years this has so developed that it is now easy to recognize certain fundamentals of classification, which may be stated as follows:

i. Classification is a human device. Nature, somehow, made the animals; but man made zoology, and in particular man made taxonomy, the science of classification. He made it, and is still making it, solely to suit his own convenience. This is a fact which students nearly always, and more experienced systematists too often, overlook. They seem to work as if there must exist somewhere a perfect or ideal and correct classification of all animals and of every group of animals. This is not so. When Nature made the animals she did not make a couple of thousand false scorpions and then throw them at Man with the challenge, 'Now sort out that lot.' On the contrary, the classification man produces must be, and can only be, the scheme or arrangement that is most helpful in certain circumstances.

ii. Classifications are nearly always based on external characteristics, easily seen and compared, a direct consequence of their practical utility. Collections of animals are always being made in all parts of the world, and these are usually sent to specialists, who work as a rule in museums, for examination and report. Specimens already familiar can then be recognized and new discoveries, if any, can be described. The basing of classifications on outward appearance can thus be understood. It is quicker and easier to characterize an animal in a description, and to recognize an animal from a description, if attention is given only to its outside. Moreover, it may well be undesirable to dissect or to section a precious specimen of a rare animal. To do so satisfactorily may require more skill than the describer possesses, and even if it is done it may reveal details which future taxonomists may not have the power to discern. Yet from the outset it is obvious that external features alone are unlikely to tell us the whole story of an animal's relationships. When the internal organs of all the species in a group have been compared, the system of their classifying may have to be changed, and this means that later workers may henceforward have the added labour of dissection to identify the material on which they are working.

iii. Classification is neither precise nor quantitative; orders and families in different classes are not equivalently distinct, because they are divisions of convenience, not of nature. A taxonomist, at work on a group of animals which particularly interests him, has no rule to tell him whether a particular order is to be divided into two or twelve or twenty families, nor whether any of the families ought to hold one genus or six. Nor need the degree and extent of the differences between two orders of mammals, for example, be the same, or even nearly the same, as the differences between two orders of insects. Decisions on these matters are, in the first instance at least, solely matters of personal opinion, and also depend in part on the number of species involved.

iv. If taxonomy is to become a scientific study of animal groupings, it must be more than a mere index, more than a list of families and genera arranged in alphabetical or historical order. Just as anyone completely ignorant of zoology would put the wolf, the fox and the dog together in a group of some kind, and the lobster, prawn and shrimp together in another, so taxonomists try to arrange their schemes and tables in a way that will recall the course of evolution. This is often expressed in the sentence, 'taxonomy is the mirror of evolution', a statement that might be more accurately phrased in the form, 'taxonomy attempts to mirror evolution'.

The results of these attempts have been only partly successful, not so much because classification and evolution are inherently diverse, but because the evolution of animal life has been a complex, heterogeneous, cosmopolitan and unsynchronized process, whereas the writing of a list of names of groupings is a very simple one.

In the easiest instances the disparity is not evident. The familiar series Fishes, Amphibia, Reptiles, Birds, Mammals, with which the most elementary student is familiar, represents the fact that among vertebrates a geological succession of at least five classes has been recognized. But even so it does not record the fact that the first mammals appeared before the first birds, and that both are descendants of reptiles, and that birds, from the point of view of instincts, behaviour, voice and structure are as highly evolved as mammals.

When a large number of groups are being considered the difficulties are multiplied. One result of this is an increased opportunity for individual opinion to claim a hearing, and to receive complete or partial acceptance or rejection. In the face of these alternatives there can be no surprise that one of the invariably admitted features of classification is that no scheme can be expected easily to attain finality.

There is, however, universal agreement as to the principles on which a classification can best be founded, the establishment of a hierarchial system, the most important feature of which is its elasticity.

Linnaeus, on whose work all later classifications have been founded, was content with four grades below the kingdom—phylum, class, order and genus: family was soon added. Increasing numbers of species and closer attention to detail in describing them first led to the introduction of the prefixes

13

super- and sub-, which, placed before any or all of the above grades, give a possible total of fifteen steps:

super-phylum	super-class	super-order
phylum	class	order
sub-phylum	sub-class	sub-order
super-family	super-genus	
family	genus	
sub-family	sub-genus	

For some taxonomic problems, however, even fifteen grades, or taxa as they are now called, have proved to be insufficient and additions or interpolations are therefore needed. Two steps have been widely adopted, namely cohort, between class and order, and tribe, between family and genus. Further, the prefix infra-, placed below sub-, increases the number of available taxa to 31.

KINGDOM
Sub-kingdom
Infra-kingdom
  Super-phylum
  PHYLUM
  Sub-phylum
  Infra-phylum
    Super-class
    CLASS
    Sub-class
    Infra-class
      Super-cohort
      COHORT
      Sub-cohort
      Infra-cohort

Super-order	Super-genus
ORDER	GENUS
Sub-order	Sub-genus
Infra-order	Infra-genus
Super-family	Super-species
FAMILY	SPECIES
Sub-family	Sub-species
Infra-family	Infra-species

In the present state of development of taxonomy this provides more steps or taxa than have yet been found to be needed.

The classifying of animals is based on homology, that is on similarities believed to be due to descent from a common ancestor. Thus we put into one class, Aves, all animals with feathered wings, but insects' wings do not qualify their owners for admission to the same class. Their wings are analogous, performing the same function, but they have no evolutionary relation to the wings of birds.

Among the higher taxa, the classes and phyla, difficulties of decision are rare. Animals with non-cellular bodies appear to form an acceptable phylum Protozoa, animals with three divisions to their bodies and three pairs of legs are Insects, those with hair and milk glands are Mammals, and so on. We are often content to assume that such classes as Insecta and Mammalia are monophyletic, that is to say they are descendants from one primaeval stock whose ancestral characteristics, though modified by the passage of time, are represented in the group today. The units of which these groups are composed are universally described as species.

The simple, unsophisticated idea that all the different 'kinds' of animals and plants are separable as distinct products of nature derives from superficial or mildly interested observation of living things. To the uninitiated it seems to be self-evident that rats and mice, dogs and cats, wasps and bees, are of different sorts, breeding true and so self-perpetuating through the ages. The notion of the distinctiveness of all species was expressed by Aristotle and Pliny; it may be deduced from the theory of Special Creation, accompanied by the idea of the fixity of species or the belief that they are unalterable, and it remained almost unquestioned until the end of the eighteenth century. But a closer study of animals has raised difficulties in accepting these opinions, and the 'species problem' is a fundamental one.

Behind the description of any species lies the assumption that all its members are almost exactly alike, that they have always presented the same appearance and that they will always do so. This is virtually a disbelief in the truth of evolution, but zoologists often forget this, feeling that evolution operates so slowly that a detailed description is justified for all practical purposes. To put this another way, the species was a static conception, and all individuals could be related to a single type. Species were monophyletic.

Common experience, however, shows that varieties often occur, varieties in colour, pattern or proportions, which do not

15

seem to justify the separation of the variant as a new species. Some of these are constant and well recognized, some are new and surprising. Increased collecting over wider and wider areas reveals many instances of closely similar species which may be found separated from each other by a natural barrier, or which may, on the other hand, overlap and interbreed. If this happens the two or more species are regarded as one, and the idea of a polyphyletic species becomes an essential.

The realization of the importance of geographical range as supplementing if not superseding structural characteristics has brought a different set of ideas to taxonomy. Each species must be recognized as having its own range of habitat and its own range of variation for each character. The solitary and precious type-specimen loses much of its former value, for the practising systematist requires a series of specimens, and even as long a series as possible from as wide an area as possible.

When these considerations and others are borne in mind it becomes clear that a short definition of the term 'species' is not possible if the definition is to have all the meaning and implication desirable. That there are many other such terms in science need scarcely be pointed out. A possible definition of 'species' according to present-day ideas is that given by Huxley:

A species is a natural unit (a) having a geographical distribution area (b) which is self-perpetuating as a group (c) which is morphologically distinguishable from other groups (d) the members of which do not normally interbreed with members of other groups.

Clearly the modern conception of species is wider than of old, and not every zoologist regards the word in the same way. The real cause of the trouble here, as in many other problems in learning, is that man has found an idea (here the idea of a natural unit) and in attempting to maintain it is faced with the difficulty of forcing other facts of nature to conform to an idea which they had never suggested.

If the idea of a species, fixed or evolving, inclusive or exclusive, requires some consideration, so also do the natures of a genus and a family.

In the simplest terms a genus is a group of related species. A species seems to be a natural unit, whereas a genus is a relatively artificial grouping. It is a collective category, which seems to imply that normally it may be expected to contain

16

more than one species, but zoologists have never agreed on an optimum size for their inventions. There are some zoologists who maintain that all the species in a genus need not show all the characteristics which were used to define the genus, an opinion which comes perilously near to making nonsense of the whole of classification.

The character of a family, an assembly of a number of related genera, is similar to that of a genus but is more inclusive. It follows that it is rare for a genus, and rarer for a species, to be moved from one family to another; also that there is a greater tendency to make sub-families than to make sub-genera. The insertion of a sub-genus may complicate the name of a species included therein. In the name Ammothea (Achelia) echinata, the word Achelia is the name of a sub-genus of the genus Ammothea, but the setting up of sub-families does not disturb the names of the species. It often makes the identification of a specimen simpler and quicker.

A question frequently asked is whether the family and the higher taxa of order, class and phylum, have an objective existence in the natural world, or whether they are purely subjective, mere creations of the systematists. The question may be answered in two ways.

The first follows from a remark made earlier in this chapter, that nature made the animals and man made systematics. In so far as this is true, it designates all taxa as subjective concepts, where not only systematics but all science itself may be said to have no existence save in the minds of the scientists.

The second answer, however, is not so sweeping. Animal species having their own characteristics, their own distribution, and an undeniable quality of survival over many millions of years, must be accepted as objective or natural. Also the verdict on a group of phyla, exhibiting the progress of life from the water, to the land and then into the air, must be an admission that such evolution is equally objective, equally the result of natural processes. What is acceptable at the bottom and at the top of this hierarchy cannot logically be denied to its middle parts, which show clearly enough the same creative design.

There is but one point to make in conclusion. Taxonomy makes an attempt to portray the progress of evolution, yet nearly all zoologists who have tried to construct a scheme of classification doing this have been forced to be content with an imperfect result. One specialist, writing of the Mollusca, has said that they are too numerous and too diverse to 'give a neat and tidy scheme of classification'. It is an anthropomorphism to expect neatness and tidiness where nature itself is neither neat nor tidy, but it must again be remembered that classification is itself the product of the neat and tidy mind of the scientists.

CHAPTER THREE

CLASSIFICATION OF THE ANIMAL KINGDOM

The only rational conclusion to the last chapter, in which the principles and practice of classification were discussed, is to apply these principles to the animals of the world. To do this with complete satisfaction, to print '*The* Classification of Animals' in a form with which every zoologist would agree, is impossible, for there is still too much ignorance, too much uncertainty and too much room for individual opinion to permit the construction of a stable system. Nevertheless, '*A* Classification of Animals' can be offered; and it may be accompanied by notes which, calling attention to some of the doubts, difficulties and divergencies, will give it a greater interest and meaning than any dogmatically issued scheme could possess.

Accordingly, the few pages that follow do not pretend to be a treatise on animal taxonomy. They deal only with phyla and classes. The result is an attempt to suggest one way in which the higher groups of animals may be arranged, with an attempt to combine convenience with phylogenetic accuracy. Arrangement is far more easily attained than accuracy.

Two points emerge. Some inclusive and familiar names are truly phylogenetic, like Amphibia and Echinodermata; some others are purely convenient or descriptive, like 'Entomostraca' for all the crustaceans that were not obviously Malacostraca; or like 'Sporozoa', for the parasitic Protozoa. Most of the latter have been abandoned in the same way as has 'Invertebrata', and like it may be retained adjectively, as descriptive labels.

Second, it may be observed that the best-known and most frequently studied groups, like the birds and the insects, are those in which the difficulties seem to be the greatest and the divergencies of opinion most acute. This would be regrettable were it not that it emphasizes the tremendous difficulty in arranging the animals of our world in a logical, acceptable, and practicable classification.

Any scheme of classification should explain itself by the inclusion of short descriptions of the important characteristics of each group that it includes. These short descriptions, properly known as diagnoses, and at one time always written in Latin, are separated from the Classification printed here, and will be found as convenient introductory paragraphs to the different chapters in Parts 3 and 4.

SUB-KINGDOM PROTOZOA

Phylum Protozoa
    Class Mastigophora (Flagellata)
        Sub-class Phytomastigina
        Sub-class Zoomastigina
    Class Rhizopoda (Sarcodina)
    Class Actinopoda
    Class Sporozoa (Telosporidia)
        Sub-class Gregarinomorpha
        Sub-class Coccidiomorpha
    Class Cnidosporidia
    Class Ciliata (Ciliophora)
        Sub-class Holotricha
        Sub-class Spirotricha

The above is a conventional classification of the Protozoa, but it is one with which many protozoologists may be inclined to quarrel, for the Protozoa have been described as representatives of a stage of organization, rather than as a naturally coherent group. Thus Sandon has recently offered a scheme in which the names of groups that cannot satisfactorily be regarded as natural, in the evolutionary sense that all are descendants of a single ancestor, are printed in inverted commas. The result, in shortened form, is:

'Class' Flagellata
    Six orders and two 'orders'
'Class' Sarcodina
    One order and five 'orders'
'Class' Sporozoa; really two distinct classes, Telosporidia, probably derived from flagellate ancestors, and Cnidosporidia, which may have had a metazoon ancestry
'Class' Ciliata
    Eleven orders
'Class' Opalinata

There could scarcely be a better example of what Sandon has well called 'the untidy profusion of nature at its worst'.

20

SUB-KINGDOM MESOZOA

The two orders Orthonectides and Dicyemides, sometimes placed in this group, were formerly classed as Platyhelminthes which had undergone degeneration.

SUB-KINGDOM PARAZOA

Phylum Porifera (Spongiida)
   Class Nuda
   Class Gelatinosa

This is a recent classification suggested by Burton, and is markedly different from the traditional division of the sponges into three classes:
Calcarea, Hexactinellida, Demospongiae

SUB-KINGDOM METAZOA (Eumetazoa)

Infra-kingdom Diploblastica
  Super-phylum Coelenterata
    Phylum Cnidaria
      Class Hydrozoa
      Class Scyphozoa
      Class Anthozoa
    Phylum Ctenophora
      Class Tentacula
      Class Nuda
Infra-kingdom Triploblastica
  Super-phylum Acoelomata
    Phylum Platyhelminthes
      Class Turbellaria
      Class Trematoda
      Class Cestoda

There is an alternative arrangement for the Platyhelminthes which differs considerably from the above established system. Its author is Prof. J. Baer, and he introduces six classes:

Turbellaria (unchanged)	Cestoaria	
Temnocephaloidea	Cestoda	
Monogenea	Trematoda	

Phylum Nemertini
   Class Anopla
   Class Enopla

A phylum named Aschelminthes is a large, comprehensive and rather heterogeneous group included by some authors in the six classes following, here given their traditional status as separate phyla:

Super-phylum Aschelminthes
    Phylum Rotifera
    Phylum Gastrotricha
    Phylum Kinorhynchia (Echinoderida)
    Phylum Nematomorpha
    Phylum Nematoda
      Class Phasmidia
      Class Aphasmidia
    Phylum Priapuloidea
    Phylum Sipunculoidea
    Phylum Echiuroidea

The three last-named phyla form the phylum Gephyrea of some authorities. The last two have also been placed as classes of the Annelida.

Super-phylum Coelomata
    Phylum Polyzoa (Bryozoa)
        Class Phylactolaemata
        Class Gymnolaemata
    Phylum Phoronida
    Phylum Brachiopoda
        Class Inarticulata
        Class Articulata
    Phylum Mollusca
        Class Aplacophora
        Class Polyplacophora
        Class Monoplacophora
        Class Gastropoda
          Sub-class Prosobranchia (Streptoneura)
          Sub-class Opisthobranchia
          Sub-class Pulmonata
        Class Scaphopoda
        Class Lamellibranchiata (Pelecypoda)
        Class Cephalopoda

The first two classes named above are sometimes regarded as sub-classes of a class Amphineura:

Phylum Annelida
   Class Polychaeta
   Class Oligochaeta
   Class Hirudinea
   Class Archiannelida
Phylum Onychophora
   Class Onychophora
Phylum Arthropoda
  Sub-phylum Mandibulata
    Super-class Myriapoda
     Class Diplopoda
      Sub-class Pselaphognatha
      Sub-class Chilognatha
     Class Chilopoda
      Sub-class Epimorpha
      Sub-class Anamorpha
     Class Pauropoda
     Class Symphyla
     Class Insecta (Hexapoda)
      Sub-class Apterygota
      Sub-class Pterygota
       Infra-class Palaeoptera
       Infra-class Neoptera
        Super-order Polyneoptera
        Super-order Paraneoptera
        Super-order Oligoneoptera.

The names used above are recent replacements for the almost equivalent system based on their metamorphoses and the degree of changes associated with these. Very nearly:

    Apterygota = Ametabola
    Pterygota   = Metabola, in which the
    Palaeoptera = Hemimetabola
    Polyneoptera = Heterometabola
    Oligoneoptera = Holometabola.
Alternatively, the
   Hemi- and
   Heterometabola = Exopterygota
    Oligoneoptera = Endopterygota,

this last pair of alternatives recalling the modes of origin of the wings. Some entomologists introduced the name Paurometabola for those of the Heterometabola whose eggs hatched as nymphs on land and not as naiads in water.

Class Crustacea
    Sub-class Branchiopoda
    Sub-class Ostracoda
    Sub-class Copepoda
    Sub-class Mystacocarida
    Sub-class Branchiura
    Sub-class Cirripedia
    Sub-class Malacostraca

Sub-phylum Pycnogonida
    Class Pantopoda

Sub-phylum Chelicerata
    Class Merostomata
    Class Arachnida
        Sub-class Caulogastra
        Sub-class Latigastra
    Class Pentastomida
    Class Tardigrada

Phylum Chaetognatha
Phylum Pogonophora
Phylum Echinodermata

Sub-phylum Pelmatozoa
    Class Crinoidea
Sub-phylum Eleutherozoa

    Class Holothuroidea
    Class Echinoidea
Super-class Stelleroidea
    Class Asteroidea
    Class Ophiuroidea

The earliest Asteroidea seem to be ancestral to both the later Asteroidea and also the Ophiuroidea, and they have sometimes been separated as a sub-class Somasteroidea.

Phylum Chordata
    Sub-phylum Acraniata
              Class Hemichordata
              Class Enteropneusta
              Class Pterobranchia
       Infra-phylum Urochordata (Tunicata)
              Class Ascidiacea
              Class Thaliacea
              Class Larvacea
       Infra-phylum Cephalochordata
              Class Cephalochordata
    Sub-phylum Craniata (Vertebrata)
       Infra-phylum Agnatha
              Class Cyclostomata
       Infra-phylum Gnathostomata
          Super-class Anamnia
              Class Chondrichthyes (Selachii, Elasmo-
                  branchii)
              Class Osteichthyes
                  Sub-class Actinopterygii (Neopterygii)
                  Sub-class Crossopterygii
              Class Amphibia
          Super-class Amniota
              Class Reptilia
              Class Aves
              Class Mammalia
                  Sub-class Prototheria
                  Sub-class Metatheria
                  Sub-class Eutheria
                      Infra-class Unguiculata
                      Infra-class Glires
                      Infra-class Mutica
                      Infra-class Ferungulata

The foregoing classification has been given in the form of a mere catalogue or index of names. Clearly to possess true value to a student each name should be followed (if space allowed) by a description of characters, sufficient to explain the differences between all the groups mentioned. But the distinguishing characters are too important to be neglected; they are therefore to be found in Parts 3 and 4, brief summaries of the nature of the groups under consideration. These résumés are generally known as diagnoses or diagnostic descriptions; they need to be carefully constructed, because of their value in making clear the essential character of a group of animals and enabling one group to be easily compared with another.

CHAPTER FOUR

ZOOLOGICAL NOMENCLATURE

From the earliest times men have found that the giving of names provided the readiest means of distinguishing between the many different things of which they wished to speak, and this included the different kinds of animals in which they were interested. With the great increase in the number of species known, nomenclature becomes an important part of zoology, with its own principles, its own problems, its own enthusiasts.

For many years the principles have been in the care of an official body, the International Commission for Zoological Nomenclature, whose latest set of rules and recommendations was published in 1961. All who seek for guidance among the obscurer details of animals' names should consult its pages.

The general zoologist, however, is more concerned with using names than with inventing them, and there are several basic conventions with which all should be familiar, and which cannot be too widely known or too clearly understood and appreciated.

The first and fundamental idea is that each species of animal should be known by one name only, the game given to it when it was first adequately described in a recognized publication. In this respect the word 'first' means in or after 1758, the year in which appeared the tenth edition of Linne's *Systema Naturae*.

The type of nomenclature is known as binominal, that is to say, the name of a species consists of two words: the most familiar example is no doubt Homo sapiens, the zoological name that man has chosen for his own species. The first of these two words is the generic name or the name of the genus; the second is the specific name or the name of the species; and the two together form the binomen.

It is to be noticed that the names are written in Latin, a habit sometimes defended as one that makes the name intelligible to zoologists all over the world. While this is true, it is really only a lucky result of the fact that in former times all intellectual work was written in that language and biological nomenclature is a relic, a survival of the middle ages. Many of the generic names, especially those of the larger and more familiar animals, are the Latin words by which these creatures were known in ancient Rome. Elephas, Equus and Canis are examples. But there are more genera than there are suitable names in the Latin dictionary, so that zoologists have had to invent names, which must be 'in Latinized form'. Examples are Drosophila, Monocystis and Oryctolagus.

The specific name is one of four types. Most often it is an adjective qualifying the generic name, as in Hydra viridis, the green Hydra. Sometimes it is a second noun in opposition with the first, as in Boa constrictor. Alternatively, it may be a noun in the genitive case, either singular, for example Peiris brassicae, or the plural, as in Demodex folliculorum, while often it is the name of a person whose association with the animal is thus perpetuated. An example is Sorex granti, which shows that such names are put in the genitive case, formed by the addition of -i, or, if feminine, -ae, to the ordinary name.

There follows on this the desirability that the simple rules of Latin grammar should be followed; for instance a genus in the feminine must be followed by a specific name also in the feminine, so that, if the name of a genus changes from one gender to another the specific name must be altered appropriately. When Amphioxus (masculine) was changed to Branchiostoma (neuter) the specific name lanceolatus changed to lanceolatum.

On many occasions in the past, zoologists, either from ignorance or carelessness, have not followed this or some similar rule, and it has been sardonically said that since scientists seldom write good English it is unrealistic to expect them to write good Latin. This may be true, but it is answered by saying that it is not unreasonable to expect them to learn enough Latin to enable them to avoid follies and imperfections unworthy of zoology. All the instruction needed can be found in Appendix D in the *International Code* of 1961.

An animal's binomen is followed by an addition which is often omitted and is not, officially, an integral part of it. This is the name of the author who first bestowed the specific name upon it. This often prevents mistakes in identification, and its value is increased if the date of publication is added. Thus the full name of the edible frog is Rana esculenta Linnaeus, 1758.

Two points should be mentioned. The first is that there is no comma between the specific name and the name of the author, though a comma precedes the date. The second is that the name of the author is not necessarily printed in parentheses. Both these rules are too often disregarded.

The parentheses round the name of an author are not a suggestion that the name is unimportant; they are an indication that the name of the genus is not that under which the author first described the species. For example, Phalangium morio Fabricius becomes, on the breaking up of the too large genus, Mitopus morio (Fabricius).

These facts, and others which may be omitted from an introductory treatment, show clearly that onomatography, or the correct writing of animals' names, is not a detail, but demands some care and attention from all zoologists. Although the full name of a species may consist of two words plus author and date, there are so many ways in which a name may, in fact, be found to have been written that it is both interesting and informative to look at a list of possible variants. I have chosen as a suitable example the name of the common garden spider which is well known to everyone:

1. Araneus diadematus Clerck, 1757	Correct name, in full
2. Araneus diadematus Clerck	Correct name, abbreviated
3. Araneus diadematus	
4. A, diadematus	
5. Aranea diademata Linné, 1758	Name used by Linnaeus
6. Epeira diadema Walckenaer, 1837	Name used by Walckenaer
7. The garden spider	Common name in Britain
8. La porte-croix	Common name in France
9. La crox de St Denis	
10. Die Kreuzspinne	Common name in Germany

It may be added that 1 and 2 are valid. This is because they conform to all the accepted rules of nomenclature, the addition or omission of the date being voluntary.

Forms 5 and 6 are legitimate, which means that they were originally used in conjunction with an adequate published description, but they are now invalid because of the abandoning of the generic names Aranea and Epeira.

The International Code as at present constituted has two main objects in addition to the promotion of the correct writing of names; these are the securing of priority and stability, two characters which are unhappily at times in opposition to one another.

Priority, the permanent use of the first valid name, is partly a basic principle, partly a matter of personal justice. It has the great advantage of being entirely logical, and if it had always been followed it would have led to a complete stability in all matters of nomenclature.

However, if its application is to be retrospective or 'back-dated', it may lead, and it often has led, to the discovery that a familiar name, long established and widely used, was not in historical fact the first name under which an organism was described. If a familiar name is for this reason to be replaced, the inconvenience and confusion caused are often greater than the logical advantage. In consequence there is established a list of preserved names or *nomina conservanda*, which, though they are not in accordance with the letter of priority, are more in accord with the spirit of stability and common sense.

It cannot be too widely known that every problem of nomenclature likely to arise is certain to find its method of solution in the International Code of Zoological Nomenclature. This Code was revised as a result of the recommendations of the International Congress of Zoology, held in London in July 1958. It is readily obtainable and it should be the duty of every zoologist to acquaint himself with its recommendations. Any activity of man that is carried on in all parts of the world, whether academic, athletic or political, must be governed by some degree of mutual discipline, and in zoological nomenclature there is no excuse for any individual who is unacquainted with current practice. A study of the Code shows clearly enough that nomenclature, as it is now known, is not a very simple, empirical matter. Unsuspected doubts and difficulties appear whenever a name is called in question, and these can only be resolved by consulting the Code. To omit this does not contribute to the progress of zoology, but is more than likely to raise further difficulties later.

PART THREE
INVERTEBRATE ZOOLOGY

CHAPTER FIVE

PROTOZOA

Animals of microscopic size, the bodies of which usually contain one nucleus, though sometimes two are present. Their bodies are not divided into cells. Holophytic as well as holozoic nutrition occurs.

There are four, or perhaps five, classes:

FLAGELLATA OR MASTIGOPHORA

Protozoa with one or two flagella. In this class are many organisms which possess chromatophores and carry out photosynthesis, so that the class is a heterogeneous one, with several orders considered as either plants or animals.

RHIZOPODA OR SARCODINA

Protozoa which move by pseudopodia of varying forms, marine, freshwater, terrestrial or parasitic. Their bodies are naked or are coated with a skeleton of silica or lime.

CILIATA OR CILIOPHORA

Protozoa which move by the beating of their cilia. A homogeneous or monophyletic class.

SPOROZOA

Parasitic Protozoa of at least two distinct types, joined in a loose, heterogeneous class.

OPALINATA

Parasitic Protozoa, in which the body is covered by cilia in longitudinal rows. Two or more nuclei, similar in appearance. Endozoic in the rectum of Amphibia, etc.

The Protozoa, which have been intensively studied ever since the invention of the microscope, form one of the most interesting, most baffling and most important groups of animals. Their small size and their apparent simplicity of structure when examined under any but the most efficient microscopes, suggest that they are most closely allied to the earliest living organisms that the world has produced. Zoologists have thus hoped to find in the study of Protozoa some clues to the fundamental mystery, the origin of life.

This hope, and others, have been disappointed. The Protozoa may well have been longer on the earth than creatures of any other kind, but in that immeasurable period of time they have evolved within the limits of their structures, have spread all over the globe, and are now an elaborate, numerous, very successful and very intriguing group.

Their size has helped in their distribution. Drifting in the slightest current, carried even in the movement of a water-film, they travel easily. To this must be added the power, shown by so many, of encysting themselves in a hard protective coat, and so, like grains of dust, being blown about by every breeze. Further, they have produced a high proportion of parasitic species, with all the elaboration of life history and all the opportunities for dispersal that this mode of life affords. Among the parasites are those responsible for some of the most virulent of human diseases.

All this has been achieved by bodies that are primitive in the sense that they are not divisible into separate parts, each with its special function. Usually the body of a Protozoon contains one nucleus, sometimes two, but any larger number is uncommon in conditions of normal health. Because of this dependence on only one nucleus the Protozoa are described as non-cellular animals. Some scientists prefer to think of them as unicellular, a difference which depends on no more than the meaning attached to the word 'cell'. In either case, the fact remains that no nucleus in a protozoon is specifically responsible for, or in charge of, one particular metabolic function. This characteristic clearly distinguishes the Protozoa from all other animals, and is expressed by their being included in a sub-kingdom of their own, also called Protozoa, co-extensive with the phylum and contrasted with all other animals in other sub-kingdoms.

FLAGELLATA (MASTIGOPHORA)

The flagellated Protozoa which form this class are divisible into two sub-classes, the Phytomastigina and the Zoomastigina. The Phytomastigina includes those that, because they contain chlorophyll or some other pigments in their bodies, are able to make their own food in the same way as plants, that is, they are holophytic in their mode of nutrition. The Zoomastigina are the colourless, essentially animal-like flagellates.

Since there are many who argue that photosynthesis, by which carbon dioxide and water are converted into carbohydrates, must have enabled plant-like organisms to appear in the world before animals that typically feed on them, the opinion is heard that Phytomastigina must be among the most primitive forms of animal life, and that the colourless flagellates have arisen from them. The opposite opinion also finds support: the chemical changes involved in photosynthesis, it is said, are so complex that it is unlikely that they would have come into existence in the presumably simple, inorganic environment of the time. Therefore animal-like organisms, able to nourish themselves on matter of inorganic origin, must have been the first to appear.

These are matters of speculation: we have no real knowledge either of the condition of the world when life first appeared, or of the process of its creation. Attention may be turned to the structural characteristic of this class, the flagellum, for this is an organ of locomotion which is universally found among both animals and plants. It is now known that the structure of a flagellum is fundamentally the same as that of a cilium, that both contain a paired axis, an axoneme, in the centre, surrounded by a ring of nine more apparently double fibres. These stiffening fibres meet within the body of the organism at a spherical blepharoplast (basal granule or kinetosome) which may be connected to the nucleus by a rhizoplast.

A flagellum is usually regarded as the first type of organelle evolved to produce locomotion in living things. Supporting this belief there is its widespread occurrence among Bacteria, Algae and Protozoa, its occasional existence in parts of other organisms such as Sponges and Coelenterata, and its almost invariable use as the means by which both plant antherozoids and animal spermatozoa are able to swim. This is certainly a point of great significance.

Undoubtedly, too, the flagella have undergone a degree of evolution; they have improved and they have sometimes changed their function. Some flagella arise from the anterior end of the organism and drag it through the water; others from the posterior end, whence they propel it. Some organisms have two flagella, in which case one may be active in advance while the other trails; or one may act as an anchor. Others again, as in Trypanosoma, form the outer edge of a thin film of proto-plasm which is used to propel the animal like an undulating membrane or even the uninterrupted dorsal fin of a fish.

It is evident that the energy which the flagellum needs to perform its function is in part derived from the body of the organism and in part is generated along the length of the flagellum itself.

Among the genera that belong to the Flagellata, the most exciting is undoubtedly Euglena, which contains about 150 species, varying from ·012 mm to ·5 mm in length. The most familiar species is Euglena viridis (Fig. 1), to be found in clean

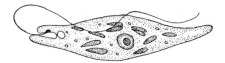

Fig. 1. Euglena viridis

fresh water, where it swims actively while carrying out photo-synthesis with the help of its chlorophyll. When it was first discovered, much argument was devoted to a determination of its real nature—was it an animal or a plant? Its chlorophyll, its cell wall, its power of storing the starchlike carbohydrate paramylum, are all plant characteristics; while its mobility, its red eye-spot, and its possession of a 'gullet' believed to ingest solid food particles, were animal characteristics. The truth is that there is no reason for supposing that an organism must be wholly animal or wholly plant, and that at the early stage of evolution represented by Euglena there was no such thing as a plant, no such thing as an animal.

Indeed, the chloroplasts found in the genus Euglena differ in number and form in different species. Some species have no chromatophores; some have haematochromes, which occupy the centre of the cell in dull light, making the creature look green, and the periphery in bright light, making it look red. Moreover, if kept in the dark some species lose their pigment and are thus dependent on other sources than photosynthesis for their food. The species E. gracilis has been known to produce colourless strains during a period of rapid binary fission in which some of the daughters missed their share of chromato-phores. When this occurs there is no recognizable difference between the offspring of Euglena and the species Astasia longa.

In the same way, some species of Chlamydomonas, as traditionally Plant No. 1 as Amoeba is traditionally Animal No. 1, may give rise to an organism indistinguishable from a species of Polytoma.

When it is added that Euglena morpha has three flagella and that it lives in the gut of a tadpole where both green and colourless varieties are to be found—that Chrysamoeba carries a yellow pigment with the help of which it can feed holophytically until it loses its flagellum, moves by pseudopodia and lives holozooically—that Noctiluca, a spherical form almost a millimeter in diameter, produces light and is one of the causes

Fig. 2. Trypanosoma

of the phosphorescence of the ocean, it becomes clear that the Flagellata admirably illustrate the very high degree of structural complexity and intricacy of habits and life history which the 'simplest' animals, the Protozoa, have evolved.

Among the Zoomastigina are to be numbered Trypanosoma, Trichomonas, Trichonympha and Opalina.

The genus Trypanosoma (Fig. 2) contains three species which, dangerous human parasites, are the causes of 'sleeping sickness' or trypanosomiasis. In Africa, T. gamiense and T. rhodesiense occur; T. cruxi inhabits South America and Mexico. The African species are morphologically indistinguishable in man, but are distributed by different species of tsetse fly, while T. rhodesiense and T. brucei are indistinguishable in the flies but affect man and animals differently.

The African species swim freely in the blood stream after a man has been bitten by Glossina palpalis or Glossina pallidipes. In the gut of the fly the parasite undergoes a series of changes, during part of which it loses its flagellum, the Leishmania phase. Later, in the crithidial phase, it swims to the pharynx and thence to the salivary glands, from which it reaches the next human victim. The American species is found inside the cells of various organs. It is transmitted by the bug Triatoma infestans and several other vectors.

Many other species of the genus are continuously present in a non-pathogenic condition in many wild animals, such as rats, and especially ungulates, and the difficulties of controlling them are very great.

Of the other genera mentioned above, Trichomonas contains the species T. vaginalis, which is apparently unable to live outside the body of the human female but, though almost universally distributed, it appears to do no harm, while its relative, T. foetus, is a cause of abortion in cattle. The species T. tenax lives in the human mouth and T. hominis in the colon, and both are harmless symbiotes.

Trichonympha collaris is a similar inhabitant of the gut of the white ant or termite, where it is a useful, indeed an essential, colleague. Wood, which is the chief diet of white ants, contains very little protein or sugar and a great deal of cellulose, which is almost indigestible. Trichonympha ingests small particles of wood, converts it into more sugar than it needs, and makes the excess available to its hosts.

RHIZOPODA

The class Rhizopoda contains all Protozoa that move by means of pseudopodia, and the orders into which it is divided differ considerably from each other. Most conspicuously the differences are found in the pseudopodia themselves.

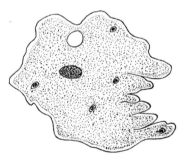

Fig. 3. Amoeba

In the order Amoebina they are usually blunt in form, of no great length in comparison with the rest of the animal, and do not anastomose, or rejoin, at a distance from it. To this order belongs the genus Amoeba (Fig. 3) whose name is as widely known and whose mode of life as familiar as, perhaps, any other invertebrate.

The common belief that Amoeba proteus represents the simplest form that animal life can take is based in large part on the apparent simplicity of its body when viewed through a moderately efficient microscope: it is a belief that calls for some consideration. The production of pseudopodia has often been called a very primitive method of progression, yet the formation of a pseudopodium is so complex that its correct explanation is still uncertain, and its mechanism seems to produce a combination of rolling and gliding very difficult to understand.

The structurelessness of the body of Amoeba is deceptive. Critical and experimental observation provide evidence of a fore-and-aft organization by which 'forward' movement is established and reversal, though possible, is rare.

The engulfing of food particles in a food vacuole again appears to be simple, but while some species surround their food by two pseudopodia others creep over the particle and almost pick it up. The different species have acquired quite different dietary habits; some are almost omnivorous while others have very definite preferences. For example, Vampyrella will only consume Spirogyra. A special process, known as pinacytosis, enables an Amoeba to drink. The surface wrinkles, the wrinkles close to form tubes the lips of which gulp in drops of water. A drink may last half an hour, may be acceptable if the water contains proteins and refused if it contains sugar.

Although Amoeba retains the primitive feature of binary fission, it shows an elaborate life history, involving the formation of spores, their hatching, and the growth of the young Amoeba. This is ultimately followed by obvious senescence.

Allied to Amoeba is the genus Entamoeba, the chief characteristic of which is that in general its species live within the bodies of other animals, including man. The species E. coli is universally present in the human intestine, where it has reached a state of equilibrium such that neither species suffers disadvantage. It can therefore scarcely be called a parasite. At intervals it encysts, the cysts are evacuated and their survival depends on their reaching a second host. In the interval the nucleus has divided three times, and when the cyst is dissolved the liberated Entamoeba divides to form eight daughters, each with one nucleus.

Its relative, E. gingivalis, inhabits the teeth and gums. It is probably harmless, and though more numerous in cases of pyorrhoea has not been proved to be a cause of this trouble.

By contrast E. histolytica is a formidable parasite. It, too, is

very widely distributed and in many of its hosts lives unnoticed in the colon, feeding on bacteria and other matter. On
occasions and in certain ill-defined circumstances, it changes its
way of life. It secretes an enzyme, cytolase, which destroys the
epithelial cells of the intestine, allowing the Entamoeba to enter
the tissues and feed there. The result is the serious disease
known as entamoebic dysentery, which is often fatal.

Other Protozoa commonly found in man are Endolimax
nana, Iodamoeba butschlii and Dientamoeba fragilis. The
bodies of these Protozoa and their allies are naked and unprotected by anything other than the superficial sarcolemma;
other orders, however, secrete a skeleton around themselves.

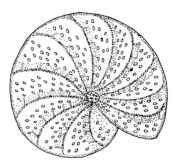

Fig. 4. Shell of Polystomella

The Heliozoa are found in fresh water. Their skeletons are
made of silica and a spicule of silica occupies the middle of each
of their radiating pseudopodia. These straight pseudopodia
make the animal resemble a sun with its rays, but they can
bend enough to push food particles towards the centre, where
they are engulfed.

The Foraminifera are marine Protozoa with skeletons or
shells of chalk. Their pseudopodia are long and thin and project
through minute holes in the shell, at a short distance from
which they may join. Many species 'outgrow' their shells and
form another chamber, and this process repeated produces
spirally coiled shells (Fig. 4) of a type familiar in many Mollusca.
A most interesting feature of the Foraminifera is that after
death their shells sink to the ocean bed and, lying there, cover
vast areas with a deposit known as Globigerina ooze. This is the
source of the chalk from which the cliffs of Dover and elsewhere

have been made, representing unnumbered millions of generations of individuals of the genus Globigerina.

The Radiolaria, like the Heliozoa, extract silica from the water and make their shells from it. These flinty remains are found as Radiolarian ooze at greater depths of the ocean where the chalk of the Foraminifera has been dissolved by the greater concentration of carbon dioxide at a high pressure. Both these orders include many species of quite fantastic beauty, first portrayed by Ernst Haeckel in his *Kunstformen der Natur*.

CILIATA

The Ciliata form a much more uniform class than either of those just described. Their bodies are covered, wholly or in part, by cilia, by means of which they move rapidly through the water. Their structure shows a greater complexity than is to be found in any other protozoan class, emphasizing very clearly the lesson that a small body is not necessarily a simple one.

Every biologist knows something of the genus Paramecium (Fig. 5), one or more species of which are so easily to be found

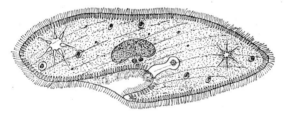

FIG. 5. Paramecium caudatum

in infusions of vegetable matter. Attention may be directed to the permanent shape, a contrast to Amoeba, with a rounded bow and a pointed stern, responsible for the old-fashioned name of the Slipper Animalcule. Within there are two nuclei, a large meganucleus concerned with day to day activities, and a smaller micronucleus of reproductive function. There are two contractile vacuoles, supplied by radiating channels and pulsating alternately as they carry out their function of preserving osmotic equilibrium between the protoplasm and the surrounding water. The cilia are arranged in rows; they beat with a slow forward motion and, stiffened, with a rapid backstroke. They are connected at their bases by visible strands called myonemes, and their rhythmical action drives the animal forwards, rotating about its axis as it goes.

In addition there are short rod-like bodies known as trichocysts to be seen in the ectoplasm. They can be shot out as long threads when the animal is irritated, but their function in its life is uncertain.

The class includes freshwater, marine and entozooic forms, these last being the species that live within the bodies of other animals. Here they may be harmless, harmful or beneficial. The most familiar of the first group are Balantidium entozoon and Opalina ranarum, both of which are easily to be found in the rectum of a frog. The latter, which with its allies is sometimes put in a separate class, the Opalinata, is one of the largest of the Protozoa and may exceed a millimetre in length. Allied to the species of Balantidium found in the frog is B. coli, a common inhabitant of the gut of the pig. It occasionally finds its way into the gut of the human species, where it may be quite unnoticed until, like E. histolytica, it suddenly attacks the tissues and produces balantidial dysentery.

An example of a valuable ciliate is Botrichia prostoma, an inhabitant of the gut of the sheep. It is supposed to absorb a proportion of the carbohydrate in grass which would otherwise be decomposed by bacteria. When, lower in the intestines, the Botrichia is digested the rescued carbohydrate helps to nourish the sheep.

The Ciliata are of great zoological interest by reason of the process known as conjugation, which they exhibit.

Essentially the life cycle of a Ciliate consists of growth followed by binary fission, followed by growth, and so on, the implication of which is that binary fission is the only means of increasing the number of individuals. Occurring, as it may, two or three times a day, it is efficient from an arithmetical point of view, but it is not a method by which fresh combinations of genes and chromosomes can be created; and the advantage of this to the survival of a species is generally recognized.

The process of conjugation, which has excited wonderment for many years, has often been described and photographed, and has been supposed to be nature's method of avoiding that peculiar formation of imperfect organisms known as depression. Whether this is its main function or not, its result is the formation of a new meganucleus from the fractions of two micronuclei.

Conjugation may occupy from fifteen to forty-eight hours. Two individuals approach and lie with their gullets in contact. Their meganuclei break up and disappear; their micronuclei enlarge and divide twice, and three of the quarters disappear. The remaining one divides unequally into two pronuclei, large and small. The latter are mobile and each moves across to the other partner and fuses with its sedentary pronucleus. The conjugants now separate. Normally each is rejuvenated; alternatively the operation may be fatal.

In a modified process, called autogamy, the same nuclear changes occur inside a solitary individual.

SPOROZOA

Into the class Sporozoa are put all the parasitic Protozoa which, at some time in their lives, migrate in the form of spores from one host to another. It contains a wide diversity of species, which may conveniently be separated into the Teleosporidia and the Cnidosporidia (or Neosporidia). The first of these two sub-classes contains the orders Gregarinidea and Coccidiomorpha; the second is of small importance.

Among the Gregarinidea, which are nearly all parasites of invertebrates, are those genera that are of interest because they have apparently reached a state of equilibrium with their hosts, and the harm that parasites are supposed to do has been reduced to negligible proportions. By far the most familiar to the student of zoology is Monocystis, two species of which, M. lumbrici and M. magna, are to be found in the vesiculae seminales of almost every earthworm that is dissected.

The life history of this parasite includes two phases. The trophozoite (adult or feeding stage) is an elongated organism a few millimeters long, with a single nucleus and contractile fibrils (myonemes) running through its body. These give it an ability to make slow, wriggling movements, so characteristic that they have been given the name of gregarine movements.

The various stages in its reproductive process can usually be distinguished in a single smear of the contents of a vesicula seminalis. In the beginning, two individuals, gametocytes, approach one another and secrete an association cyst around themselves. In this their nuclei divide several times, and the fragments, with a small amount of surrounding protoplasm, escape and fuse in pairs. The zygotes enclose themselves in an

43

easily recognizable boat-shaped cyst, the sporocyst or pseudona-
vicella, in which it divides into eight sporozoites. By some
means, as yet unknown, these sporozoites enter a second worm;
probably the transference follows the death of the original host
and liberation of the pseudonavicellae into the soil; possibly
they reach the soil in the faeces of a bird or frog that has eaten
the earthworm. Once in this new host the sporozoites make
their way to the vesiculae seminales, where they can be seen
covered with the tails of developing spermatozoa.

In the sub-order Haemosporidia the outstanding genus is
Plasmodium, four species of which are responsible for malaria.
These are Plasmodium vivax, P. malariae, P. falciparum and
P. ovale.

The parasite enters the blood stream of a human being in the
form of a spindle-shaped sporozoite which was previously living
in the salivary gland of a mosquito. In thousands they enter the
liver, feed on its cells, and each undergoes multiple fission
(schizogony) producing merozoites. In less than a week there
may be from one to four thousand merozoites from each
original sporozoite. These merozoites now enter the red corpuscles, in which the process of schizogony is continued until
the corpuscle bursts and the new generation of merozoites is
set free to infect more corpuscles and repeat schizogony. It is
thus clear that the number of parasites may soon exceed
millions.

Ultimately the merozoites cease to undergo fission and grow
into gametocytes which can develop no farther in the human
body. But when the blood containing them is sucked in by
another mosquito they survive, and in its gut they produce
male and female gametes. Fertilization may occur within half
an hour of the mosquito's meal. The zygote, called an ookinete,
penetrates the gut of the insect and forms an oocyst like a
blister on the outer surface of the stomach. In this cyst repeated
divisions give rise to sporozoites, which, set free into the
haemocoele, make their way to the salivary glands, en route for
a human victim.

Malaria is one of the most formidable of all human diseases.
Its control depended on Ross's discovery of the part played by
the mosquito as vector of the parasite, and is the best historical
example of the fact that zoological knowledge of the animals
involved is a pre-requisite of success by the medical profession.

CHAPTER SIX

PORIFERA

Multicellular animals in which the cells are arranged in two or three layers around a central cavity. There is no nervous system, and the cells influence each other only by chemical or mechanical activity. The cells are of a collared-flagellate type. Many sponges reproduce asexually by budding; gametes are also emitted and the zygote develops into a ciliated, free-swimming larva. The colony is usually strengthened by a skeleton of spongin, sometimes with spicules of chalk or silica.

There are three classes:

CALCAREA

Sponges in which the skeleton is composed solely of calcareous spicules.

HEXACTINELLIDA

Deep-sea sponges in which the skeleton consists of six-rayed siliceous spicules.

DEMOSPONGIAE

Sponges in which the spongin may contain no silica, or silica not in the form of six-rayed spicules.

The essential form of a sponge is that of a vase or hollow vessel, radially symmetrical or irregular in shape. The central cavity or paragaster is surrounded by three layers of cells, the innermost of which, the gastric layer, contains the cells that are almost peculiar to this phylum. These are the collared flagellate cells or choanocytes (Fig. 6). The middle layer or mesenchyme is composed of separated cells set in a jelly-like medium, into which the cells have secreted spicules of a chalky or siliceous nature. These constitute the supporting skeleton of the colony, so that the cells that produce them are called the skeletogenous cells. Outside the whole there is an epithelium-like layer of flat cells known as pinacocytes.

The choanocytes resemble the stalked Protozoa of the family Choanoflagellidae. Their flagella which, without correlation or mutual dependence cause currents of water to enter the many pores on the surface of the colony and to escape through the main orifice of the paragaster, are continuously active. Sponges are the only organisms in which the main orifice is exhalent. The water brings in the food which the cells require, either in the form of small organisms that are trapped on the collars of

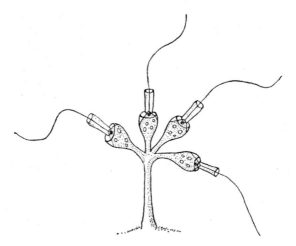

Fig. 6. A Choanoflagellate organism

the cells, or simply as nutritive matter in solution. The collar is not a smooth cone but is made from a number of divergent, overlapping elements.

In a few sponges the central gastric cavity is a simple unbranched hollow; in others it is extended by lateral branches, and in others again these branches themselves branch.

The spicules in the mesenchyme do not articulate or even necessarily touch each other. In some sponges the spicules are held by fibres of a horny material, spongin, in others a fibrous skeleton of spongin is present, but spicules are absent.

Sponges can reproduce asexually by budding, and they show so marked a power of regeneration that any fragment torn or cut from a colony is able to add the features that it has lost while the parent colony replaces the lost portion.

The phylum Porifera, or the sub-kingdom of the sponges, occupies a unique, almost a phenomenal position among the animal organisms of the world. Because of their collared flagellate cells they seem to be derivable from a protozoon ancestor; because the separate cells of a colony neither influence each other nor co-operate in any way for the benefit of the whole, they are clearly distinct from any of the animals known as Metazoa.

A suggestion which at once springs to the mind is that sponges represent an intermediate stage between the cellular and the non-cellular, and that they even provide evidence of the way in which this great evolutionary step was taken. Is this wishful thinking or is it a tenable opinion? If it is tenable, then the Porifera may be expected to show resemblances to both the Protozoa and the Metazoa.

The resemblances between the Protozoa and the Porifera are neither numerous nor convincing. The existence of choano-flagellate cells may be followed by reference to the organism Proterospongia, described by Saville Kent in 1880. It consisted of a flat plate of cells in which the outermost were flagellated while the inner ones were amoeboid, and it is therefore not surprising that it has been considered to be a primitive sponge showing protozoon characters. More recently, however, Proterospongia has been thought to be no more than a stage in the development of a fresh water sponge.

There are more resemblances between the sponges and the Cnidaria. Both are sedentary, aquatic organisms with a tendency to branching colonies, and both can reproduce asexually and sexually. On the other hand, the Porifera never possess nematocysts and have neither nerve nor myo-epithelial cells. The problem is not simplified by our inability to say whether the Porifera or the Cnidaria appeared on the earth first, nor by the undoubted fact that in some ways the sponges are specialized organisms.

CHAPTER SEVEN

COELENTERATA

> Multicellular aquatic animals in which the body
> consists of two layers of cells surrounding a
> single cavity, the coelenteron, with one opening
> to the exterior. Radial symmetry is the rule, and
> the most characteristic feature is the offensive-
> defensive weapon, the thread cell or cnido-
> blast. A simple nervous system is present. Some
> genera are simple, others colonial. There are
> two types or forms of individual, the sedentary
> hydroid and the motile medusoid, both occurring
> in some classes.

The name Coelenterata may now be used for a super-phylum,
of which the Cnidaria and the Ctenophora are the component
phyla. The Cnidaria are those that possess the thread cells,
which are absent from the Ctenophora. There are three classes:

HYDROZOA

Cnidaria in which the polyp and the medusoid forms typically occur in the life cycle.

SCYPHOZOA (jelly-fish)

Cnidaria in which the medusa is the main or the only form.

ANTHOZOA (sea anemones and corals)

Cnidaria in which the polyp is the only form.

The most important feature of any member of the Coelenterata is the arrangement of the cells in two layers, the ectoderm and the endoderm (Fig. 7). Since each of these layers
is normally only one cell deep it resembles an epithelium of a
higher animal, and for this reason the Coelenterata have been
described as animals which show the cell layer or epithelial
grade of organization. The implication is that this represents
a stage of evolution intermediate between the cell grade (Protozoa) and the organ grade (Triploblastica).

This type of body is accompanied by other features always

found in association with it. The two layers are separated by a non-cellular, gelatinous layer, the mesogloea, into which cells are sometimes found to have wandered. The central hollow, the coelenteron, unites the functions of the coelom and the enteron of higher animals, and its one opening, the mouth, with secretory, digestive cells in it, is a contrast to the dominant exhalent opening found in sponges.

Many of the Cnidaria are sessile animals and show two common accompaniments of the sessile habit: they are radially symmetrical and their bodies may branch, forming a colonial

FIG. 7. Transverse section of a Coelenterate

organism. Sessile animals require special means of dispersal, and among the Cnidaria dispersal is the function of a characteristic product, the medusa. The function of feeding the individual or the colony is carried out by the polyps, which occur at the ends of branches; hence there is a division of labour between the two forms.

The shape of the polyp is that of an oral cone with the mouth at its apex, surrounded by a ring of tentacles. The number of tentacles varies considerably and is not even constant within a species. They are the organs by which prey is captured.

The medusa is always compared to a short-handled umbrella; the handle of the umbrella ends at the mouth and the tentacles are distributed round the circular edge. A system of radiating canals runs from the centre of the edge.

Two features of the medusa are important. The first is the existence of eight sense organs, or statocysts, equally spaced round the circumference of the umbrella. Each contains a granule of calcium carbonate, which is secreted by a special cell in connection with the nerve ring. It is supposed that they help to orientate the medusa with respect to gravity. The second feature is that it reproduces by sexual methods. The gonads, male or female, are situated in the middle of four of the radiating canals. They shed their gametes into the water, where fertilization occurs. The zygote develops into a characteristic planula larva, which, swimming by means of its cilia, helps to spread the race.

Cnidaria, which are solely carnivorous, capture their prey with the help of nematocysts. These are bladder-like objects containing a coiled thread which, appropriately stimulated, is

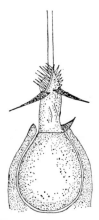

FIG. 8. Nematocyst

shot out, piercing, poisoning or entangling the victim. Frequently described and commonly regarded as characteristic of the Cnidaria, they are found, none the less, among Protozoa of the genus Polykrikos. The nematocysts of the Cnidaria are of at least three different kinds, which have been designated as penetrants, glutinants and volvants. Each is found in a special ectoderm cell known as a cnidoblast or nematoblast, provided with a sensory projection, the cnidocil (Fig. 8). When the cnidocil is stimulated by the passing of an organism, pressure seems to be exerted on the nematocyst, everting the thread. Cnidoblasts are especially numerous on the tentacles, and when they have captured a victim it is carried to the mouth which may be stretched open to receive it.

Under the action of digestive enzymes secreted by endoderm cells the food disintegrates in the coelenteron and particles are engulfed by the cells of the endoderm, which assimilate them. Ectoderm cells depend for nourishment on diffusion of digested products from the endoderm.

This account of feeding implies the existence of some kind of nervous coordination. Nerve cells, polyhedral in shape and each with its central nucleus, are distributed all over the mesogloea, below the ectoderm. Each cell gives off a number of processes or nerve fibres, which meet and join similar processes from other nerve cells. In this way a nerve net is formed all over the animal or colony.

HYDROZOA

The general features of Cnidaria as outlined above are perhaps best shown in the class Hydrozoa, where perhaps the genus Obelia is the most useful illustration.

Obelia geniculata occurs in the form of short, white thread-like growths on rocks and seaweeds, to the surface of which it sticks by a set of hollow tubes, its hydrorhiza. From these rise the branching stems of the colony. The stem consists of a solid coenosarc enclosed in a chitinous perisarc. The branches end in bell-mouthed hydranths, which are enlargements of the coenosarc, enclosing the oral cone and its surrounding ring of solid tentacles, engaged in catching food.

In the axils of the hydranths appear modified buds, whose function is the production of medusae. Each such bud consists of a gonotheca shaped like a tall Roman urn, with a rod-like blastostyle in the centre. The medusae develop on the blastostyle and as they mature are liberated into the water. The medusae of Obelia differ from those of most genera in that they lack the velum, or narrow horizontal shelf that runs round the inside of the bell. It is composed of contractile cells and at its base is a double nerve ring.

The zygotes which arise from the gametes liberated by medusae develop first into a hollow blastula, and then into a solid, pear-shaped larva, termed a planula. This swims by means of cilia until it settles on its broader end and produces a coelenteron by a separation of endoderm cells, while a mouth and tentacles appear at the top.

As an illustration of the differences found among the Hydrozoa, the species Bougainvillea fructuosa may be mentioned. Here the perisarc, cut short below the polyp head, does not form a bell-like hydrotheca; there is no blastostyle and the medusae form directly on the coenosarc, from which they break away.

An example of a solitary polyp is seen in Corymorpha. In this genus there is no hydrorhiza; the lower end of the coenosarc is buried in the sand and bears a second ring of tentacles a little below the apical polyp-head.

Again, another kind of colony is found in Hydractinia, which grows on the whelk shell occupied by a hermit crab. It coats

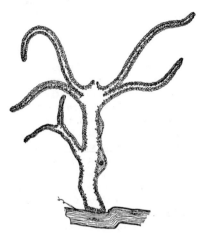

Fig. 9. Hydra, as seen in section

the surface of the shell with a mat of hydrorhiza, from which the polyps spring. Some of these are nutritive, some reproductive. They bear clusters of medusa-like objects which never separate from the colony. The ova develop while on the polyp into free-swimming larvae.

In other Hydrozoa a gradual degeneration of the medusoid phase is to be seen. Thus in Eudendrium the gametocytes travel along the coenosarc towards a developing gonotheca, from which they ultimately escape. A continuance of this leads to the loss of the gonotheca and the development of the gonads on the body of the polyp.

This is the condition found in Hydra (Fig. 9), the usual elementary type of Cnidaria. The three British species, Chlorohydra viridissima, Hydra oligactis and H. attenuata, are thus rather atypical of the group. They live in fresh water, a lucky fact which makes them more readily available for study and which seems, by comparison with other animals, to be responsible for the omission of the free-swimming medusoid phase.

Finally, there exist permanently free Hydrozoa in which there is no trace of the medusoid. The best-known example is Physalia, the Portuguese man o' war, whose beautiful blue body is a formidable inhabitant of the warmer seas. The blue colour is that of a bladder-like float or pneumatophore, which acts like a sail as the colony is driven by the wind. The tentacles hang into the water below, where the nematocysts capture food for the colony. Among them are reproductive zooids, from which gametes escape. The stinging power of Physalia is sufficient to make it unpleasant to bathers; the species occasionally appears off the coasts of Devon or Cornwall where it has arrived on the Gulf Stream.

ANTHOZOA

In this class there are no medusae. The polyps, solitary or colonial, are familiar to all as the sea anemones or as corals, and the differences between a sea anemone and a solitary member of the Hydrozoa are superficial rather than fundamental.

The body of an anthozoon is a short cylinder, usually attached to a rock surface by its basal disc and moving only rarely. Circles or spirals of hollow tentacles surround the mouth, which is a slit-like opening, ciliated at its ends. The tentacles are well furnished with nematocysts, and small animals captured by them are pushed into the mouth. The upper part of the body is strongly muscular, and if the tentacles are roughly treated they are withdrawn and the whole body changes from a beautiful flower-like object to a smooth ovoid mass. This is familiarly seen when the anemones of our rocks are left exposed by low tide.

Of the differences between this class and the last, the first that may be mentioned is that part of the oral disc has been pushed in at the mouth, forming a short gullet which leads to the gastric cavity. This cavity is divided internally by perpendicular sheets of tissue known as mesenteries. They grow inwards from the body wall towards the gullet, with which they sometimes make contact. There are eight mesenteries in some families, six in others, and some multiple of six in others. They secrete digestive enzymes and the general effect is to increase the absorptive surface of the interior.

On the mesenteries also are ridges of reproductive cells, producing ova or spermatozoa. These escape through the mouth and unite in the sea, where they develop into planula larvae.

Coral-forming polyps such as Madreporaria are essentially anemone-like creatures whose ectoderm cells secrete an external shell of calcium carbonate. A few of the stone-corals are to be found in the cooler waters of the north Atlantic, but most of the group are confined to the warmer seas. A solitary coral has, like an anemone, a ring of tentacles round a circular plate: in the middle of this is the mouth, and the plate shows the edges of thin sheets radiating from it, the digestive mesenteries. The whole polyp head is resting in a chalky or stony cup which it has made itself.

Both solitary and branching corals occur; some of the branching species may be more than a metre across and contain thousands of polyp heads. The growth of the colony ceases when it reaches the surface of the water and forms the basis of a coral reef. These sometimes take the form of a shore reef, at no great distance from the coast of the mainland, or a barrier reef which encircles an island, or an atoll, enclosing a lagoon. Coral islands have always held a place of their own in romantic literature, and to a zoologist this is justified by their almost unbelievable diversity of colour and form, enhanced by the multitude of fishes, sponges and coelenterates forming an association like nothing else on earth.

SCYPHOZOA

The Scyphozoa or jelly-fish contrast with the Anthozoa since they have subordinated the hydroid and emphasized the medusoid phase. Thus the common jelly-fish of our shores, Aurelia aurita, which is found all over the world, is recognizably like an enormous Obelia medusa with several obvious differences.

The manubrium is larger and the mouth at its end is star-shaped with four pointed lips, in each of which there are cilia. These drive food and water into the gullet and thence to the central stomach or gastric cavity. Four chambers or pouches lead off the stomach, and on their walls are the horseshoe-shaped gonads. From the stomach there also run four single and four branched radiating canals, leading to the circular canal which communicates with the hollow tentacles.

The sexes are separate and the gonads are of a purple-red colour. Ova and spermatozoa pass into the gastric cavity and escape through the mouth. The zygote develops into a characteristic planula larva, which for a while swims by means of cilia. It then becomes sessile and later splits by a number of horizontal divisions into saucer-like free-swimming ephyra, which grow into the medusa form of the adults.

Most Scyphozoa resemble Aurelia in all essentials, their differences being shown in such features as the shape of the bell and the length of the tentacles. Only one genus, Craspidacusta, is found in fresh water, and one species, Pelagia noctiluca, is phosporescent.

Jelly-fish are peculiar among all the animals of the world in the proportion of water which their bodies contain. This amounts in some specimens to 90 per cent of their weight, and a dead Aurelia cast up on the sand leaves hardly any noticeable residue when the sun has been shining on it for two or three hours.

The customary view, expressed or implied in many elementary zoology texts, is that the Coelenterata are naturally the lowest or the first phylum of the Metazoa, and that this opinion is supported by their possession of two layers of cells.

The idea that any phylum of invertebrates is necessarily and intrinsically lower or more primitive than any other must be derived from a conception of the process of evolution in which the various types or forms or patterns of animals' bodies have followed each other in orderly succession, each presumably in some ways an improvement on its predecessors. If this is a true picture of the past, then it must follow that one phylum or other group of Metazoa must have been the first in the field and must therefore be regarded as the primitive stock from which all other groups have arisen. But is this true?

The truth is that it is hard to find any indisputable evidence which suggests that the Cnidaria are the favourite candidates for this post. They are in competition with at least three other phyla, the Porifera, Ctenophora and Platyhelminthes, all of which have had some support for their claims; and in the present state of our knowledge it is not possible to come to any indisputable decision.

CHAPTER EIGHT

CTENOPHORA

Mobile, marine diploblastica, which swim by
combs or ctenidia composed of cilia fused in rows.
There are no nematocysts, their function being
performed by characteristic lasso-cells. The body
shows bilateral symmetry and is permeated by
a system of canals from the enteron to the
exterior.

The sea gooseberries or comb jellies which compose this
interesting group are transparent marine organisms which
float or drift near the surface of the sea, or swim somewhat
feebly with the help of a peculiar arrangement of cilia (Fig. 10).

Fig. 10. Hormiphora plumosa, a Ctenophore

A typical Ctenophore, such as Hormiphora (Fig. 10) or
Pleurobrachia, is an ovoid or pear-shaped animal with eight
bands of cilia, often compared to lines of longitude on a globe.
Each band consists of short transverse rows of cilia rising from
a common base and so resembling a comb, to which the name
of the phylum is due. As the cilia beat they produce a display
of prismatic colours which give Ctenophora a remarkable and

characteristic beauty. In addition to this, some species are luminous at night.

The body has a mouth in the centre of one pole, leading to a straight gut or gastro-vascular cavity, which gives off eight caeca towards the comb bands. Indigestible residues escape through the mouth. At the opposite pole there is a sense organ or statocyst which contains grains of chalk in varying contact with projections within. This implies the existence of a nervous system, the centre of which is a mass of nerve cells round the statocyst. From this centre radiates a nerve net similar to that of the Cnidaria, but consisting chiefly of eight cords which lie under the comb bands. It has been found that if the nerve centre is removed the cilia of the combs are no longer synchronized and their beating becomes irregular. If one nerve cord is cut, the cilia below the cut cease to keep time with those above.

The reproductive organs of Ctenophora are found on the sides of the diverticula from the gut. Both ovaries and testes occur in the same individual, which is therefore hermaphrodite. The gametes escape through the mouth into the water, where they meet and self-fertilization is probable. The zygote develops into a larva which, in at least one genus, Gastrodes, is a planula and thus supports the idea that Ctenophora are allies of the Cnidaria.

Near the mouth is a second feature of Ctenophora, a pair of long tentacles. These arise from small pits into which they can be withdrawn: they are the organs with which the animal catches its food. Nematocysts do not take a share in this, for one of the characteristics of the Ctenophora is the absence of these cells. Their place is taken by sticky or lasso cells, which cause the tentacles as they sweep through the water to capture small crustaceans and even tiny fish. The tentacles then carry the food to the mouth in much the same way as the tentacles of Hydra.

The species Hormiphora plumosa and Pleurobrachia pileus, which are common off the British coasts, are of the typical globular form, but other shapes of the body are found in the phylum. The girdle of Venus, or Cestus veneris, is a long ribbon-like animal, about two inches wide and maybe a yard long. It floats as part of the plankton in the Mediterranean, while Coeloplana crawls on the bottom and recalls the mode of life of a planarian.

Clearly the relationship between the Cnidaria and the

Ctenophora is neither close nor obvious. Some members of the latter show a body form which can be compared to that of a medusa, for there is a single layer of cells on the outside of the body, a similar endoderm lines the gut and between the two there is a gelatinous mesenchyme. This contains a few amoeboid cells and some long contractile fibres which represent muscle. Such resemblances are, however, almost as misleading as helpful. The Ctenophora are an isolated group, with certain relations to the more familiar Coelenterates and also to the Platyhelminthes, leading to the probability that all three phyla are the descendants of a common ancestor.

CHAPTER NINE

PLATYHELMINTHES

Acoelomate triploblastica, in which the alimentary canal has a single opening. Excretory, osmo-regulatory organs are present in the form of solenocytes. A nervous system is present. There is a complex reproductive system, which is hermaphrodite in most genera. Both free-living and parasitic forms exist.

There are three classes:

TURBELLARIA

Free-living Platyhelminthes covered with cilia and with no trace of division into proglottides.

TREMATODA (flukes)

Parasites with an oval or leaf-shaped body with mouth and alimentary canal, and without repeated segmentation of the body.

CESTODA (tapeworms)

Internal parasites without mouth or alimentary canal, the body consisting of a variable number of parts or proglottides.

The phylum Platyhelminthes shows us for the first time a body in which the constituent cells may be divided into three groups. In addition to the ectoderm and endoderm of the Coelenterata there is an intermediate mass of a cellular nature, the mesoderm (Fig. 11). This third layer is found in all the remaining (higher) phyla, and in most animals it forms the chief components of their bodies.

In the Platyhelminthes the mesoderm is still in a relatively primitive condition. Much of it consists of mesenchyme, or space-filling 'packing tissue', comparable to the parenchyma of plants. It contains, however, some characteristic muscle fibres, produced by special myoblast cells, and its other cells help in the transport of digested food. But the essential feature of the mesenchyme is that it is solid, in contrast to the mesoderm proper of the higher phyla, where it lines an important cavity, the coelom.

In addition, the free-living Platyhelminthes introduce a feature so familiar that it may well pass unnoticed. The Turbellaria are all animals with a definite head at a permanently anterior end, with a corresponding posterior end, with distinguishable dorsal and ventral surfaces, and so with a left and

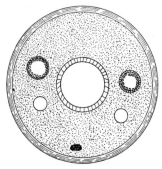

FIG. 11. Section of a tripoplastic animal showing gut, excretory and gonadial canals, and solid ventral nerve cord

a right side. This bilateral symmetry is retained throughout almost the whole of the rest of the animal kingdom.

With it there appears a new system, the excretory or perhaps more accurately the osmo-regulatory organs. In Platyhelminthes this consists of a network of tubes on each side of the body, opening to the exterior through a number of pores. Many of the branching tubes end blindly at characteristic hollow cells, called flame-bulbs. In the hollow of these cells there is a group of active flagella whose constant movement drives a current of fluid along the tube. Their appearance is that of a flickering flame, hence the alternative name for solenocyte is flame-cell. In small aquatic animals like Turbellaria most of the excretory matter escapes by diffusion, and the probable value of the solenocytes is to control the concentrations of the body fluids.

A Platyhelminth has no heart, blood or vascular system. It has, however, a nervous system, which sets a pattern followed, with minor modifications, by most of the other invertebrate phyla. In a Planarian there is to be found a pair of nerve ganglia ('brain') near the anterior end, and from these two, four or more nerve fibres run back to the tail. They have no ganglia, but a number of cross-connections preserve the form of a nervous network. The system is an advance on that of the Coelenterata, and it is retained in the parasitic classes where external stimuli must be rare and insignificant.

Nearly all Platyhelminthes are hermaphrodite. There are generally a number of testes, hollow objects secreting spermatozoa into their interiors, which lead by vasa efferentia to two vasa deferentia. These may dilate to serve as vesiculae seminales before joining to form a muscular, eversible penis. The ovary is usually median and is in close connection with a yolk gland, and from it the oviduct, with a swelling or receptaculum seminis, runs to the genital atrium with which the penis also communicates. Cross-fertilization appears to be the rule.

TURBELLARIA

Members of this class are normally aquatic, free-living and ciliated; they never show strobilization, and seldom develop suckers. There are six orders, four of which deserve mention.

ACOELA

This order is composed of small worms, seldom more than a tenth of an inch in length; it includes the genus Convoluta, so often mentioned as an example of symbiosis. The striking internal feature is the existence of what is usually described as a solid alimentary canal. This sounds like a contradiction in terms; actually a food particle is received into a mass of loose endoderm cells, in which it is digested. There is a rudimentary nervous system and a sense organ which resembles that found in the Ctenophora. In several respects the order lies between the Ctenophora and the other orders of the Turbellaria.

RHABDOCOELA

This order contains marine and freshwater worms from a tenth to a quarter of an inch long. They have a more efficient nervous system, a hollow alimentary canal and excretory organs in the form of flame-bulbs. The genus Microstomum is of unusual interest because of the presence of nematocysts in its ectoderm. These have come from Cnidaria which the rhabdocoele has eaten: unexploded, they are adopted by special cells in the gut and carried to the ectoderm, where they are detained and used in defence against predators.

61

This order contains aquatic and land forms, some of which attain a foot in length. The familiar British species, Dendrocoelum luteum and Planaria lugubris (Fig. 12), belong to this order which takes its name from the three branches of its alimentary canal. The mouth is in the middle of the lower surface of the body, at the end of a tubular eversible pharynx.

These genera have attracted attention for two reasons. The first is their apparent indifference to starvation. After the normal store of nourishment has been exhausted various

Fig. 12. A Planarion

organs of the body are absorbed and in consequence the animal shrinks in size. This may continue until it is only a sixth of its usual length, but when food is again available the lost tissues are soon replaced.

Further, Planaria possesses an extraordinary power of regeneration. If a specimen is divided, almost any piece will grow into a complete individual: if the head is split longitudinally each half will become a complete head. Grafts from other individuals can be easily implanted and remarkable specimens manufactured artificially in this way.

POLYCLADIDA

This order is wholly marine and some of its species are six inches long. Their eggs hatch into objects known as Miller's larvae, with eight ciliated lobes, again recalling the Ctenophora.

TREMATODA

This class is a wholly parasitic group, in which the liver fluke of the sheep is the most familiar. There are two orders:

MONOGENEA OR HETEROCOTYLEA

Parasitic in one host only.

DIGENEA OR MALACOTYLEA

Parasitic in two or more hosts.

The species of the Monogenea have a characteristic posterior sucker, which is often subdivided, their paired excretory orifices are placed near the anterior end of the body and the male and female systems have separate openings.

The Monogenea are nearly all gill parasites of fishes, where, since they often live singly, self-fertilization must be the rule. By one of the freaks of science, the species most often seen by zoologists working in laboratories is an exceptional one, an internal parasite found in the bladder of the frog. This is Polystoma interrimum, recognizable at once from its peculiar environment and its group of six posterior suckers. It is a hermaphrodite, in which cross-fertilization by copulation occurs. The eggs are laid in the bladder, pass out through the cloaca, and hatch as ciliated larvae with a large posterior sucker. These larvae can survive only if within twenty-four hours they encounter a tadpole to whose gills they attach themselves. They now lose their cilia and make their way to the bladder where they may remain for as long as three years. If, however, a larva finds a tadpole with external gills to which it can become fixed it develops much more rapidly, matures and lays eggs within a few weeks preceding its death, which is inevitable when the tadpole undergoes metamorphosis.

The Digenea include all the more familiar flukes, of which the liver fluke of the sheep, Fasciola hepatica, is the most often studied as an example of the group.

The adults usually live in such places as the bile duct of a sheep or the alimentary canal of a pig or a human being. Each is of a familiar leaf-like shape, with one anterior sucker surrounding its mouth and a posterior sucker at a short distance behind it. There is no anus. The interior is filled with mesenchyme cells in which the various organs lie. The mouth leads into a short pharynx, dividing into two longitudinal intestines, which branch copiously on each side of the body. There are numerous flame-bulbs, like those already described, whose ducts open into an excretory duct with a median orifice near the anterior end. Respiration is anaerobic, energy being derived from the fermentation of glycogen in the cells.

The liver fluke is hermaphrodite, with very complex reproductive organs. Two testes, consisting of masses of branching tubes, both produce a vas deferens and the two join and lead to a vesicula seminalis. Hence an ejaculatory duct leads to a muscular cirrus, with a median orifice. The single ovary is also a branched tubular structure from which an oviduct, joined by a pair of tubes from yolk glands, leads to a wide, coiled tube, the uterus. Cross-fertilization is believed to occur.

Eggs, each surrounded by a clear shell, are laid in thousands and pass out with the faeces. The first small larva, the miracidium, hatch from each, swimming with the cilia that cover it until it meets the mollusc that is its secondary host. For each species of Trematode there is one, or perhaps two or three species of mollusc suitable for this purpose, and if the miracidium does not find one in a few hours it dies.

If it is fortunate it bores into the body of the snail and turns into a sporocyst. The sporocyst is active; it feeds on the tissues of its host and makes its way to the heart, while within it a third larva, the redia, is being formed. Each sporocyst produces a number of rediae from each of which the next, fourth, larval stage, the cerceria, is formed. This is a tadpole-like organism which escapes from the snail and swims in the water until it settles on a plant where it encysts. It re-enters its primary host when the host eats the plant to which the cyst is fixed.

As a pest of sheep, Fasciola is most formidable on damp pastures and especially in early autumn. It can be controlled by dosing the individual sheep or by spraying with copper sulphate the field inhabited by the snails.

A similar parasite of man is Fasciolopsis buski, the adults of which live in the intestine and whose larvae can make use of three or four species of snails. It should be added that the life history of some Trematoda involves three hosts. For example, the adults of Opisthorchis sinensis are parasitic in the liver of men, dogs and cats, badgers and pigs. The eggs are swallowed by a snail, whence the cerceria swim to the third host, a fish. It regains the first host when the fish is eaten raw or imperfectly cooked.

A very different type of trematode infection is produced by members of the family Schizostomatidae. The blood fluke is the common name of the species Schizostoma (Bilharzia) haematobium whose adults live in the abdominal veins. The male and female are unusually closely allied, the male holding the female in a special gynaecophoral groove. The eggs are laid in the blood vessels of the bladder and each, being provided with a sharp spine, pierces the bladder wall, enters it and escapes with the urine. The piercing of the bladder is the chief cause of injury to the host. The miracidia find snails, and the cerceria, which have long forked tails, swim in the water until the primary host washes in or drinks it. In either event the parasite enters the blood vessels and makes its way to the abdominal region.

CESTODA

Tapeworms, which are in general more familiar than liver flukes, are very different in appearance. The adults, which live in the alimentary canals of vertebrates, consist of a minute spherical head or scolex which, by means of hooks or suckers or both, maintains a firm hold on the intestinal wall. Behind the head is a short neck, the function of which is the continuous production of segments, properly called proglottides, of which the rest of the tapeworm is made. In this way, the older

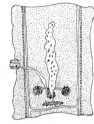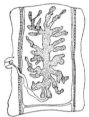

FIG. 13. Proglottides of Taenia: proximal, median, distal

proglottides are farther from the head, the younger ones nearer to it. The total number reached depends on the species. The small Taenia echinococcus consists of only three, while Dibothriocephalus may attain a total of four thousand and a length of eighty feet.

The tapeworm has neither gut nor cilia, for it lives a stationary life immersed in predigested food. A nerve collar gives off two nerves which run the length of the animal; so too do two excretory canals. These lie near the side of each proglottis, united by a transverse vessel near the hinder margin. They are fed by numerous flame-bulbs lying in the mesenchyme.

In each proglottis there are male and female reproductive organs, the male developing in the younger stages and the female alone occupying the older (Fig. 13). Numerous testes scattered throughout the proglottis are connected by vasa deferentia to an ejaculatory organ, which in most species opens at the side. Older proglottides contain more obvious female organs. There are two ovaries connected to a yolk gland and a

65

shell gland: their united ducts give off a median cylindrical sac, the uterus, and continue as the vagina to the common lateral orifice.

Normally a host contains but a single parasite, so that fertilization must be effected between two proglottides belonging to the same individual. The fertilized ova pass into the uterus which, as the remaining sex organs disappear, branches and grows until it occupies the whole of the proglottis and contains some hundreds of thousands of eggs.

Ultimately a time comes when, singly or in small groups, the terminal proglottides break off and pass out of their host. The time of their survival may well be correlated with the habits of the intermediate host, for example, the Davainea proglottides are released towards evening when slugs and snails are feeding.

The free proglottis has often a slight power of movement; that of Taenia saginata, for example, can crawl a little, shedding eggs as it goes, and a notable example is Raillietina cesticilus which moves rapidly to the faecal surface because its secondary host, a beetle, is not copraphagous. In the stomach of the secondary host the wall of the proglottis is digested, and from the ova set free by dissolving of the egg shell emerges the hexacanth embryo or oncosphere. This, by its six hooks, bores its way into a blood vessel and is carried to a muscle where it comes to rest. It loses its hooks and swells to form a bladder worm or cysticercus. Inside this one or more inverted scolices develop and the bladder worm undergoes no further change as long as its host lives. But, if the flesh is eaten raw or undercooked by the primary host the bile stimulates the turning outward of the new scolex, the bladder is dissolved, the head attaches itself to the intestine and starts the production of proglottides.

Many species of tapeworms are well known. Taenia saginata and T. solium are human parasites with cysticerci in the ox and the pig. The former is now almost extinct in Britain. Taenia solium lives in dogs, which pick up the cysticercus from mice, rabbits or hares: T. coenurus forms its bladder worm in the brain of the sheep, producing the disease known as staggers. Diphyllobothrium latum in the pike has a pre-cysticercoid stage in Cyclops, and a second metacysticercoid stage in sticklebacks.

It is clear that the control of tapeworm infection is possible by close examination (inspection) of meat, and by its thorough cooking. Vermifuges swallowed by the primary host may cause the breakaway of many or all the proglottides but the dislodging of the scolex is more difficult. Tapeworms in rats or mice may die in a few days, those in birds in a few months, but in man Taenia saginata has been known to remain for 55 years or longer. It is a comfort that the physical damage caused by a single tapeworm is not serious.

CHAPTER TEN

ROTIFERA

Small triploblastica with no coelom and no trace
of segmentation, but with a well-developed
alimentary canal. A circle of cilia at the anterior
end and a spiral row of cilia near the mouth
serve both for locomotion and feeding. There are
simple nervous and muscular systems. The excretory organs are solenocytes. The sexes are
separate.

The Rotifera, well known to microscopists everywhere, serve
admirably to illustrate the nature of animals at this grade of
evolutionary progress. They show resemblances to, as well as
differences from, both the Platyhelminthes and the Nematoda,
for their excretory system of flame-cells leading to a duct
which enters the gut is similar to the condition found in some
of the Trematodes, and the largely syncytial nature of the
body recalls the Nematoda. At the same time there is a very
obvious surface resemblance to some of the ciliate Protozoa.

Rotifera are nearly all freshwater animals, though a few
genera swim in the sea and some are terrestrial. They possess
a remarkable ability to resist desiccation and can recover from
a prolonged period of suspended animation when moist conditions return.

The body is variable in shape and is usually divisible into
trunk and foot, the foot frequently bifurcates. A hundred or so
nuclei are to be seen in the body, but they are not separated by
divisions to form distinct cells. Nevertheless the nuclei retain
their positions. The characteristic cilia beating rhythmically
round the mouth give an impression of a revolving wheel,
to which the name of the phylum refers.

Male rotifers are rare; they are quasi-degenerate creatures
with a small gut or no gut at all. They fertilize the females
either with the use of an intromittent organ or by penetrating
the cuticle and liberating spermatozoa in the body cavity.
The females produce eggs of two kinds: summer eggs are
thin-shelled and develop parthenogenetically, the winter eggs
are thick-shelled. They are fertilized, rest throughout the

winter, and in the spring hatch into parthenogenetic females only. Some species of Rotifera are viviparous, and in some the males are unknown.

The fascination which rotifers have long exercised over microscopists is equalled only by the zoological interest of their relations with other phyla.

They are among the smallest of the metazoa and are among the best examples of the fact, not often emphasized and seldom explained, that small size in itself brings certain advantages with it. Animals that are not much bigger than some of the Protozoa and which, yet, have cellular bodies, include some of the Turbellaria, Nematoda, Tardigrada and Acari.

Rotifera have cilia on part of their surface, and this recalls the Ciliata among the Protozoa. Is it therefore possible that Rotifera are descended from Ciliata, increasing the number of nuclei, acquiring organ systems, and so suggesting one possible way in which the Metazoa have arisen from the Protozoa? They have a hollow body without solid mesenchyme, and this distinguishes them from Platyhelminthes, but their excretory organs, which take the form of flame-bulbs, resemble those of the flatworms, and especially the arrangement in some Trematoda where they open into the alimentary canal. Again, the small number of cells and the tendency towards a syncytial condition make the Rotifera recall the Nematoda.

One of their most intriguing features is their overall resemblance to a trochosphere larva. This larva, to which reference will be made in several later chapters, is characteristic of the Annelida, the Echinodermata and the Mollusca, three large phyla with organizations considerably in advance of that of the Rotifera. Yet it may be that the ancestors of all four were at some time in the distant past closely related to each other.

There are many unanswered questions in the phylogeny of the larger animal groups, and it is fascinating to think that the clue to some of them may be hidden in the zoology of the Rotifera.

CHAPTER ELEVEN

NEMATODA

Vermiform acoelomate animals without segmentation, blood system or respiratory organs.
The body is surrounded by a tough, protein
cuticle with only the most rudimentary sense
organs. There is a well-developed nervous
system. There are no cilia. The gut is straight
and unbranched. The reproductive organs are
tubular, the ovaries paired, the testes single.

There are two classes:

APHASMIDIA

Nematoda with no caudal sensory organs and a rudimentary
excretory system.

PHASMIDIA

Nematoda with caudal sense organs and a well-developed
excretory system.

The Nematoda, or roundworms, form a group as remarkable
as any in the animal kingdom. Both free-living and parasitic
species are known; they have spread throughout the whole of
the habitable world and are often found in unimaginable
situations. Many are of considerable economic importance
since, as parasites of men, crops and herds, they cause serious
diseases.

The body of a nematode nearly always has the form of a
smooth, rounded cylinder (Fig. 14). At the fore end the mouth
is surrounded by three or more lips; the posterior end is usually
curled and in the male bears one or two penial setae. Lines run
along the middle of the back and lower surface, as well as along
each side. In the female the genital pore is about one-third of
the body length behind the mouth.

The mouth opens into a straight gut. The hinder end of the
oesophagus, the so-called pharynx, exerts a sucking action to
enable the animal to feed. The wall of the mid-gut consists of
a single layer of cells, devoid of either glands or muscles: they

absorb its contents directly so that digestion is intra-cellular, as in Hydra. The hind gut is lined with invaginated cuticle. The simplicity of the alimentary system is no doubt associated with

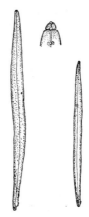

FIG. 14. Ascaris: female and male

the fact that a Nematode usually lives on the pre-digested food of its host (Fig. 15).

The nervous system is remarkably similar to that of the Platyhelminthes. There is a nerve ring or collar round the pharynx and close to it lie four ganglia, dorsal, ventral and two

FIG. 15. Ascaris, dissected

lateral. From the collar six nerves run to the sense papillae on the lips, and from the ganglia four cords run back along the length of the body, with connecting commissures at intervals. The external dorsal and ventral lines mark the tracks of these cords. The lateral lines, also mentioned above, indicate the positions of the excretory canals. No solenocytes communicate with these canals, which unite and open at a ventral excretory pore.

The hermaphrodite reproductive system is characteristic and is not complex. There are two ovaries in the form of thin tubes, gradually widening to make two oviducts followed by two uteri. These join to form a short vagina. The testis consists of one coiled tube of gradually developing spermatozoa, which opens near the tail. Cross-fertilization is the rule and the penial setae assist in the transference of the sperm, which are amoeboid and not flagellated.

Fertilized eggs are laid by the female, at the rate of tens of thousands a day, into the gut of her host and pass with the faeces to the exterior. Here they will perish if they become dry, so that they depend for their further development on the chance of being swallowed by a second host of the correct species. Normally they then hatch in its intestines, bore their way into the blood vessels and come to rest in the lungs. From here they ascend to the mouth, are swallowed again, and so reach the stomach where they mature.

This is a very brief outline of the general life cycle of a parasitic Nematode, but although most species of this phylum are parasites at some time in their lives, there are some that are freely living in the soil, and many that are parasites of man. This extreme adaptability to environments of so diverse a character is one of the most remarkable features of the phylum, and is so distinct from the comparative uniformity of surroundings that is common to most other groups that it deserves further description.

It should not be forgotten that the lives of most Nematoda include four ecdyses or moults and that after the second the larva is often protected from desiccation and other dangers by retaining the cast skin. In this form it is most likely to be active and to seek suitable surroundings for the continuation of its existence.

The subject of the diversity of a Nematode's mode of life may be considered analytically. The animal lives first as a larva, then as an adult: in either state it may be free-living or parasitic, and if it is a parasite its host may be a plant, an invertebrate animal, or a vertebrate animal. This gives a total of 16 different combinations, since each of the four possible larval conditions may be followed by one of the four possibilities for the adult. By not separating man from the vertebrates and treating the human species as a separate case, the medical importance of Nematoda is neglected. If this omission is repaired the number of combinations rises to 20. Few zoologists would be prepared to state that it is impossible to find species of Nematoda which would act as examples for every one of these alternatives, and the inclusion of man emphasizes the seriousness of the diseases that nematode infestation often produces.

It is true that all Nematoda are not dangerous. A very common parasite of man is the 'threadworm', Enterobius vermicularis, which may reach a length of half an inch and yet cause but little harm. It is most formidable in its migrations, when the movements of hundreds of tiny individuals just after hatching cause pains in the muscles. Others are not so lightly dismissed, and a truer view of the lives of Nematoda should be presented. To do this a selection of ten species has been made, intended to show briefly their range of habitat, their migratory methods, and the reasons, mainly economic, which have made them worth choosing.

ASCARIS LUMBRICOIDES

The eggs which may be deposited on plants or on the ground with excreta develop inside their shells and infect a new human host who may have taken the risk of eating unwashed and raw vegetables.

NECATOR AMERICANUS ('hookworm')

Eggs are deposited on the ground where they hatch into larvae which are free-living and which feed and grow. When mature they may burrow through the human skin, e.g. of the unshod foot of a passer-by, and so reach the blood vessels for dispersal.

TRICHINELLA SPIRALIS

The female, living in the gut of a pig or other animal, produces not eggs but larvae. These travel to the muscles, where they encyst. If this flesh is later eaten raw or under-cooked a second host is infected.

WUCHERIA BANCROFTI

The female infests the human lymph glands where she produces larvae or microfilariae. These enter the blood stream and must be sucked up by a mosquito. When this insect bites its next human victim they come out on to the skin into which they burrow. The consequent swelling of the lymph vessels and limbs is the cause of elephantiasis.

DRACUNCULUS MEDINENSIS ('guineaworm')

The female, which may be four inches long, forms ulcers near the surface of the skin, from which the larvae escape into water. Here they swim until they enter a Cyclops; the human being is re-infected by swallowing the Cyclops in drinking water.

MERMIS NIGRESCENS ('rainworm')

The adults live in the soil and their larvae enter insect larvae through the skin, living here until adult, when they return to the soil.

RHABDITIS ABERRANS

Lives freely in the soil.

ANGUILLULA ACETI

Lives in vinegar.

RHABDONEMA NIGROVENOSUM

A hermaphrodite generation lives in the lungs of the frog. Embryos escape through the mouth and mature in the water where they live until swallowed by another frog.

HETERODERA SCHACHTI ('eelworm')

The female is a parasite in the roots of tomatoes, beet, etc.; the male is free-living and fertilizes the female while her abdomen protrudes from the root. She grows to a great size and thousands of larvae escape into the soil. In some regions one-fifth of the land is so infected by eelworm that root crops cannot be cultivated successfully.

There are many species that might claim places in this list, such as Spherularia, whose adults live in the soil and whose larvae develop inside humble-bees, but the chief zoological interest of the phylum Nematoda lies in its isolation. The structure of all species is remarkably similar and they form, indeed, the ideal example of a monophyletic phylum. In three unusual respects, the lack of a coelom, the peculiar spermatozoa without flagella, and the absence of cilia, they faintly suggest arthropodan affinities. But differences between them and the Arthropoda are sufficiently obvious from the account of their structure given above, and the differences are not such as can be attributed to consequences of the parasitic habit. Resemblances to both Platyhelminthes and Rotifera are to be found, but are difficult to interpret except by describing them, insignificantly, as due to chance.

To the practising zoologists, or at least to some of them, the Nematoda have an inexplicable yet undeniable attraction. The finding of a specimen of Parascaris in a pasture where an infected horse has been at grass, or of Rhabditis under a microscope in earth in which flesh has rotted, is not an experience lightly to be forgotten, and dissection, though easy enough, vies with that of the snail in fascination. There is indeed no other group to which an enthusiastic zoologist might devote himself with greater satisfaction or greater opportunities.

CHAPTER TWELVE

ANNELIDA

Triploblastica in which the body is composed of a large number of segments covered by a thin, transparent cuticle. Below this are layers of circular and longitudinal muscles. The alimentary canal is surrounded by a fluid-filled space, the coelom, from which lead coelomoducts and nephridia (excretory organs). The nervous system consists of a pair of cerebral ganglia, joined to a solid, ventral nerve cord. The individuals are often hermaphrodite.

There are four classes:

POLYCHAETA

Annelida with a clearly defined head of several segments bearing appendages; parapodia in the form of flat extensions of

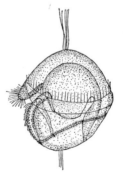

FIG. 16. Trochosphere larva

the body wall, supported by numerous setae. Usually marine and dioecious. The larva is a trochosphere (Fig. 16).

OLIGOCHAETA

Annelida with a prostomium in front of the mouth and no head appendages. There are no parapodia and the setae are fewer in number. Freshwater and terrestrial, and usually hermaphrodite, with no larva.

HIRUDINEA (leeches)

Annelida with a sucker at each end of the body. The segments are subdivided externally, and there are neither setae nor parapodia. Most are freshwater, hermaphrodite species.

ARCHIANNELIDA

Primitive, mostly marine Annelida with cilia and with imperfect segmentation and generally without setae.

The Annelida represent a clear advance in complexity of body and elaboration of organ systems.

Outwardly, the most obvious of their novelties is the division of the body into a number of segments (somites or metameres). Internally, the contents of nearly all the somites are the same; there are portions of the gut, the nerve cord, the longitudinal blood vessels, as well as a pair of nephridia and layers of muscles in each segment. Each segment or somite is separated from its neighbours by transverse septa, through which pass the organs mentioned above.

This type of segmentation, in which all segments are not only similar but are also substantially of the same age, is distinct from the strobilization of the Cestoda, where each proglottis is older than its neighbour nearer the head and younger than its neighbour farther from it.

Exceptions to the general uniformity of all segments occur near the head or in the reproductive region.

The third or middle layer of cells, the mesoderm, introduces a most important feature, known as the coelom (Fig. 17).

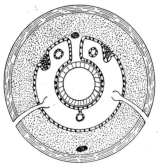

FIG. 17. Section of Annelid showing coelom

77

When the mesoderm is formed in the developing egg its cells may separate, leaving a space or schizocoele between them. Alternatively, pouches may grow out from the archenteron or primitive gut, and form spaces, an enterocoele. In either case, a space, the coelom or body cavity, is provided and in it lie the chief organ systems. Its boundaries are the splanchnic mesoderm cells, which surround the gut, and the somatic mesoderm cells, next to the epidermis.

This condition represents a great advance on the loose mesenchyme or packing cells of the Platyhelminthes or the large vacuolated cells of the Nematoda. Most simply, a perivisceral space is an advantage to any animal that moves at all rapidly.

The coelom is filled with fluid, and pressure exerted on this by the circular muscles makes the otherwise soft body of a flabby worm into one that is rigid enough to push its way through the soil. It may therefore be described as functioning like a liquid skeleton.

The coelom is always open to the exterior. In the earthworm dorsal pores exist in every somite; in the dogfish there is a pair of abdominal pores; in all animals there are ducts between the coelom and the exterior. In the Platyhelminthes, osmoregulation and to a certain extent excretion are achieved by the system of flame-bulbs already mentioned. In the Annelida these flame-bulbs have been replaced by flame-cells or solenocytes, in which the cell is smaller and the flagella finer and longer. At first they passed their excretory products into a tube known as a nephridium, which had developed from the inside outwards, and later the solenocyte disappeared and the nephridium opened directly from the coelom to the exterior by a ciliated funnel, the nephrostome. Particles of waste matter, often from the outer walls of the intestine, float in the coelomic fluid, whence they are taken into the nephrostome. Hence the coelom has an excretory function.

As well as nephridia there exist coelomoducts, developed from the outside inwards. Their original function, which they have retained, is the conveyance of the gametes to the exterior. Thus another function of the coelom becomes evident. The gonads develop in its walls, it receives the gametes, and by a special set of ducts gives them access to places where they may meet. In the reproductive somites of Lumbricus, nephridia and coelomoducts exist side by side.

The body of an annelid consists of an anterior prostomium, a number of segments or somites exhibiting metamerism, and a posterior somite bearing the anus. Generally a thin cuticle surrounds the body, and limbs or parapodia, consisting of hollow outgrowths of the body wall, are found in all somites behind the head. These parapodia are supported by chitinous chaetae or setae; they are absent from the Oligochaeta, but the setae remain.

The alimentary canal is usually a straight tube from mouth to anus. The mouth leads into a spacious buccal cavity, which is followed by a sucking pharynx. Buccal cavity and pharynx are lined by invaginations of the cuticle, which may be locally elaborated to form jaws, as in the familiar Nereis. The pharynx may lead to an enlarged crop and a muscular gizzard before becoming a straight intestine to the anus.

The blood vascular system, when present, is characteristic. Blood, containing in solution haemoglobin which possibly acts as a store of oxygen, flows forward in a dorsal vessel above the gut and backward in a ventral vessel below it. These two long vessels are connected in every somite by lateral vessels which pass through the nephridia en route. There is often also a subneural vessel. From all these longitudinal vessels there are numerous branches lying closely below the skin, which acts as a respiratory surface. There are many variations of this pattern of vessels in the different orders.

The nervous system is more constant. A pair of supra-pharyngeal ganglia, sometimes loosely called the brain, is easily seen above the pharynx. Circumpharyngeal commissures connect these to a pair of corresponding sub-pharyngeal ganglia, from which a double, solid nerve cord runs through every somite to the tail. In each somite lateral nerves to the circular and longitudinal muscles ensure co-ordination of movement. The ventral nerve cord may also contain a set of three giant fibres, whose function is apparently the rapid conduction of impulses in moments of emergency.

Most Annelida are hermaphrodite. It is the general rule among the Oligochaeta, but is less frequently to be found among the Polychaeta, where the sexes are more usually separate. In the hermaphrodite forms cross-fertilization is always the custom.

POLYCHAETA

This is an almost wholly marine class, typified by the well-known Nereis, which is often used by fishermen under the name of the ragworm. Though many polychaetes spend most of their time in burrows in the sand they can also swim freely with the help of their plate-like parapodia. They are to be found in the seas all over the world, living at all depths.

Apart from their segmentally repeated parapodia their most obvious feature is the distinct appearance of a head in the front of the body. The two characteristics of bilateral symmetry and cephalization go together, it being desirable that the same end of the body should always go first and be equipped with various sense organs (Fig. 18). These react to the nature of the

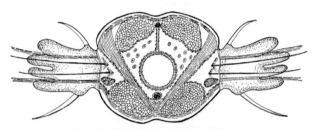

FIG. 18. Transverse section of Nereis

approaching region, so that tactile organs, organs of chemo-reception and eyes are often concentrated in the anterior segments. In Polychaeta the head is usually composed of the first four somites. The eyes may be very small, or they may be well developed and possess crystalline lenses (Fig. 19).

In some genera the somites are so different that the body may

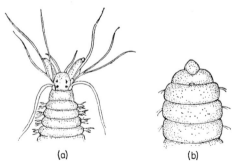

(a)   (b)

FIG. 19. Annelid heads
(a) Polychaete. (b) Oligochaete

be said to consist of a head, a thorax and a tail. Also the parapodia of some families are reduced almost to vestiges, but the setae remain.

The Polychaeta are a much more diverse class than the Oligochaeta. As well as freely swimming species, they include a number of tube makers, the tubes being made of hardened mucus or built from grains of sand and pieces of shell. The stereotropic tendency to thrust the body into a tube, even a glass tube in an aquarium, may be very strongly developed.

Among the interesting polychaetes, Aphrodite aculeata, the sea mouse, deserves mention. Its outline is an oval pointed at each end and its iridescent bristles give it an attractive glistening appearance. Another genus, Tomopteris, with long tentacles and narrow parapodia, looks outwardly more like a myriapod than an annelid, but the most diversified family is the Nereidae.

Some species of this family when they are mature undergo remarkable modification or metamorphosis, sometimes known as epitocy. The middle and hind portions lose muscles and alimentary tract, flaps appear on the parapodia, and the setae broaden and so help in swimming. There are also changes in front, including an enlarging of the tentacles and the eyes. Worms in this modified state were first described as members of a different genus, Heteronereis. In the polychaete Eunice viridis this condition coincides with a swarming or migration to the surface of the sea, where the posterior part is separated. The head reverts to its bottom-dwelling habit and regenerates its tail: the masses of egg-bearing tails which appear only at certain seasons of the moon are known as Palolo to the natives of Samoa and neighbouring islands, for whom they are a welcome source of food.

In another species of Nereis the male and female occupy the same tube and the eggs are fertilized in it. The male then eats the female and cares for the eggs and young. Among other nereids actual change of sex from male to female may occur; and there are also a few species that have become adapted to brackish or even to fresh water.

OLIGOCHAETA

This class contains the land and freshwater annelids and is universally distributed. Some of its species reach a length of four feet or more.

The general anatomy of an oligochaete is widely known because the custom of dissecting specimens forms a part of most courses of biology. The smooth cuticle, the absence of parapodia and the reduction of the setae to a much smaller number, usually eight a somite, are therefore familiar characteristics.

Most worthy of notice is the simpler structure of the head. There is a sensitive prostomium, the mouth lies ventrally behind it, and there are no eyes, no tentacles, no toothed, eversible pharynx. All these and other features suggest that an oligochaete is a less highly evolved annelid than is a polychaete,

Fig. 20. Transverse section of an oligochaete (most sections show four pairs of setae)

but the reverse is nearer the truth. The reductions just mentioned are to be regarded as adaptations, fitting the earthworm to its subterranean life in circumstances which would be intolerable to a polychaete.

This view is supported by the internal organs (Fig. 20). There is a greater elaboration of the alimentary canal, but chiefly the hermaphroditism is conspicuous and characteristic. There are elaborate arrangements that ensure internal cross-fertilization, a contrast to the more or less fortuitous external fertilization of the marine class. And the general rule of evolution is an ascent from the water to the land.

The relation between earthworms and the soil has been common knowledge ever since Darwin's famous book of 1881. There are, in Britain, between a million and three million earthworms per acre, and they move from one to 25 tons of soil an acre each year. These figures are greater than those given by Darwin.

Moreover, the idea that many worms make rich land may be looked at the other way round, for it is reciprocally true that in rich, well-cultivated soil many more earthworms are able to survive.

HIRUDINEA

The leeches, which compose this class, are different in many ways from other Annelida, differences which are partly to be correlated with their parasitic habits.

There are neither parapodia nor setae, and the somites of the body, which are usually thirty-two in number, are externally marked by grooves or rings which appear to multiply the actual degree of segmentation. But the most important external feature of a leech is a pair of suckers, one at each end of the body, formed by modification of the terminal segments. The sucker at the head is formed from the prostomium and the first two somites, with the mouth in its centre, while the tail sucker is composed of seven somites; it is directed downwards and the anus lies above it.

Internally, a leech has a central nervous system very similar to that of an oligochaete, save that the sub-pharyngeal ganglion is a composite body made of four fused ganglia. The alimentary canal is very unusual. Pharynx and oesophagus lead to a crop or midgut from which are given off a large number of lateral caeca. These are able to store a very considerable amount of blood which the leech extracts from its victim. An enzyme in the saliva prevents coagulation, and the blood is very slowly digested, a single meal sometimes lasting for weeks or months. The crop is followed by an intestine and rectum.

The excretory organs are nephridia and the reproductive organs are hermaphrodite. There are two ovaries and as many as eighteen testes, and the process of fertilization includes the unusual phenomenon of dermal impregnation. This consists of the placing by the male of a spermatophore, or packet of spermatozoa, almost anywhere on the skin of the female. The sperm penetrate the skin and make their way to the ovaries where the ova are fertilized *in situ*. A clitellum secretes a cocoon in which the young develop.

Clearly the leeches resemble the earthworms more closely than the Polychaeta. Like them, they are of worldwide distribution and occur both in fresh water and on the land. In former times they were widely used by surgeons for bleeding their patients, since the letting of a quantity of blood was held to be a cure for complaints of all sorts. At present leeches are most obtrusive in hot, damp regions. They have been a very formidable pest to armed forces in two world wars, attaching themselves to the legs of men, mules and horses.

ARCHIANNELIDA

The simplification of bodily structure which distinguishes the Oligochaeta from the Polychaeta is continued in this class. The name, the implication of which is obvious, was given when the group was believed to be the most primitive of the Annelida, but this opinion is no longer held. This illustrates the fact that while the representatives of an ancestral group may show an original simplicity of structure, the representatives of another and later group may show similar characteristics as a result of simplification.

The class Archiannelida is usually typified by Polygordius neapolitanus (Fig. 21), in which the prostomium carries a pair

Fig. 21. Polygordius neapolitanus

of tentacles on which a few setae remain. Its longitudinal muscles occupy four quadrants, and there are no circular muscles. The larva is a peculiar trochosphere, with traces of both a prostomium and posterior segments.

The genus Dinophilus, with five or six somites and rings of cilia, a reduced coelom and muscular system, recalls a small Turbellarian. There are several other genera.

Many species of this class have lost all traces of segmentation of parapodia and of setae. Their bodies are covered with cilia, a feature that may be regarded as a survival of the cilia of the larva. This retention by an adult of juvenile characters is not uncommon among animals and is known as neoteny.

CHAPTER THIRTEEN

ONYCHOPHORA

> Caterpillar-like Arthropoda, whose anterior appendages are antennae, gnathites and oral papillae, followed by many pairs of unjointed legs. Colomoducts act as excretory organs. The genital orifice lies slightly in front of the anus.

Over sixty species of these remarkably interesting animals have been described since Peripatus juliformis from the Isle of St Vincent was discovered in 1825, and was first described as a slug with legs. It is, however, not a mollusc but an animal that presents a curious mixture of annelid and arthropod characters (Fig. 22).

The species range from 1·55 to 15 mm in length; they are of various colours, orange, brown, green and even dark blue or

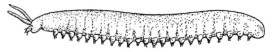

FIG. 22. Peripatus

black, with an attractive velvety surface, showing no trace of external segmentation. Dorsally, the eyes and the short annulated antennae are the most obvious features. The mouth opens on the ventral surface of the head, a circular aperture almost surrounded by a number of lips and provided with a pair of toothed jaws or gnathites. At the sides of the mouth are two oral papillae. These are short appendages, at the tips of which open slime glands, and the sticky fluid they secrete can be ejected with some force, entangling insects and other small creatures on which Peripatus preys.

All remaining appendages, which vary in number from 14 to 43 pairs, are short annulated legs ending in a pair of retractile claws. With their help the creature crawls slowly in the leaf-litter which is its normal habitat. At the base of each leg is the opening of an excretory coelomoduct.

To all outward appearance Peripatus seems to resemble an

annelid, as if one of the Polychaeta had taken to the land, converting its parapodia into legs and setae into claws. The fact that the legs, though ringed, are not really segmented or jointed is in accordance with this idea. The long worm-like body with complete metamerism, as shown by the repetition of legs, and its pair of simple eyes, are also annelidan characters.

Within, the anatomy is not so definite. Rings of circular muscles outside layers of longitudinal muscles, and the repeated excretory organs continue the list of annelid characters; but the body cavity itself is largely haemocoelic, a typical arthropodan feature. The heart is a longitudinal vessel situated dorsally, the nervous and reproductive systems are more like those of an arthropod, as also are the toothed gnathites.

It is of interest to notice that while in most species fertilization is internal, Peripatus capensis has been seen to leave the spermatozoa on the skin of the female. This 'hypodermic fertilization' is also found among some of the leeches.

The first descriptions of the organization of Peripatus regarded it as a link between the phyla Annelida and Arthropoda. Because it is rare to find evidence of close relationship between different phyla, this opinion was of particular interest. On the other hand, fossil remains tell us that the Onychophora are an extremely ancient group, in existence in Cambrian times, and that, incidentally, they have changed very little during the millions of years that have since elapsed. Present-day opinion tends to put the Onychophora in a group by themselves, with the implication that they are neither the descendants of the Annelida nor the ancestors of the Arthropoda. They may be described as a very early group of Arthropoda which retained a number of annelidan characters.

The acceptance of this conclusion supports the suggestion that at the time of origin of any grade of organization there were opportunities for the selection of a number of alternative characters. To put this idea less metaphorically, the causes of evolutionary development, whatever they were, had different results for existing animal groups. The result was a distribution of characters among the various recipients, so that one such group emerged with the characters that we now associate with Annelida, another with those of Onychophora, and yet a third with those of Arthropoda. The relationships between the three phyla could not therefore be compared to grandfather, father and son, but rather to three sons, or three grandsons, of the same ancestor.

The two other classes that have been associated with the Onychophora in the Pararthropoda may be considered shortly here.

TARDIGRADA

The Tardigrada or water bears are among the smallest and most widely distributed of the Metazoa. They seldom exceed a millimetre in length, and they are to be found in a variety of situations all over the world. One species, Macrobiotus arcticus, which occurs in arctic waters and is not uncommon in the lakes of Ross Island in the Antarctic, provides an example of bipolarity as well as an illustration of their range.

Associated with this is their apparent indifference to the conditions of their environment. One species has been recorded on dry lichen on a volcanic rock in an arid area of Mexico, others from the bottom of a fifty fathom Scottish loch, from an altitude of 2,000 feet in the Himalayas, from hot springs in Japan and from glaciers in Greenland. Several species are named in the collections from every country where these animals have been studied.

Their resistance takes two forms, described as encystment and anabiosis. In encystment, the animal is covered by a tough shell which retains moisture and in which respiration continues slowly; in anabiosis the Tardigrade shrivels and all metabolism appears to stop until conditions become more tolerable, when it absorbs water, swells and resumes active life.

The structure is peculiar. The whole body is surrounded by a cuticle which is cast at intervals, and the permanent epidermis lies below it. In many species both these are transparent, but red, brown and blue pigments occur in others. There are eight legs, usually unjointed, and attached are claws derived from the cuticle. The inside of the body is full of a transparent blood, with the internal organs lying in it and with floating corpuscles which contain a reserve of food material. The alimentary canal is a straight tube consisting of a chitin-lined mouth, a spherical suctorial pharynx, a large stomach and a rectum with a pair of lateral, probably excretory, caeca. The ovary or testis is a single organ, opening in company with an accessory gland into the rectum.

In their alimentary canal and their legs the Tardigrada recall Peripatus, and the balance of evidence seems to point to their being Arthropods of at least as primitive a character as the Onychophora.

LINGUATULIDAE

The Linguatulidae are a parasitic group whose adults live in the lungs, gut and coelom of reptiles, birds and mammals, and whose larvae are either free-living or are encysted in an intermediate host which is often a fish.

The adult is a worm-like creature from a few millimetres to as much as 9 cm long, with a body more or less divided into a short head and a longer abdomen. On the head are two pairs of chitinous hooks; the abdomen is flattened or cylindrical. The sexes are separate.

After fertilization has occurred the male is usually sneezed out of the host, but the female fixes herself in the lung tissue and lays eggs which ultimately reach the gut of the host and thence the outside world. Here they may be swallowed by the intermediate host, in whose stomach the egg membrane is dissolved, liberating the embryos. These enter the blood stream and are carried to other parts of the body, where they encyst. In the cyst they develop, undergoing ecdyses until a fully formed larva emerges, waiting until the secondary host is eaten by the primary host, in whose lungs they become adult.

CHAPTER FOURTEEN

ARTHROPODA

The phylum Arthropoda includes about three-quarters of a million species of animals whose bodies are metamerically segmented and protected by a tough, sclerotized exoskeleton. The somites are usually grouped into sets or tagmata such as form, for example, the head, thorax and abdomen of an insect. Each somite carries a hard tergite above and a hard sternite below, these two plates being connected by a softer pleural membrane. Each somite also typically carries a pair of jointed appendages, from which the name of the phylum, 'jointed feet', is derived. One or more of these pairs of appendages are modified to act as mouth parts (also called manducatory organs), and from most of the segments of many arthropods appendages are absent.

The coelom, which was so characteristic a feature of the Annelida, is much reduced in the Arthropoda, where it remains chiefly as the hollows in the reproductive and excretory systems. The general body cavity is described as haemocoelic, meaning that it is filled with blood, bathing all the organs within. The nervous system, like that of the Annelida, consists of paired cerebral ganglia followed by a solid, ventral, ganglionated nerve cord. There are no cilia and no nephridia: coelomoducts function as excretory organs and gonoducts.

The characteristics outlined in the above diagnosis seem to indicate a close, evolutionary relationship between the Annelida and the Arthropoda. True, the latter have a harder and more resistant exoskeleton than the delicate cuticle of the worms, and they have very different gonads and genital ducts, but the nervous systems in the two phyla are remarkably similar. The appendages of the Arthropoda, though characteristically jointed, are outgrowths of the body wall, just as are the parapodia of Polychaetes, and the embryonic development is of the same type.

This rather surprising set of resemblances between two groups suggests that they had a common ancestor, or that the primitive arthropods were evolved from some primitive annelid. If the ancestor of all the Arthropoda was a single type of annelid, then the phylum Arthropoda can be described as monophyletic: all its classes and orders have diverged from a single source. Alternatively, the different classes of Arthropoda may have evolved from different classes of Annelida and subsequently undergone a parallel evolution. If so, they are a polyphylectic phylum.

General opinion is not unanimous in favour of either view, but the prevailing belief shows an increasing tendency to regard as probable a polyphyletic nature in several of the traditional phyla, the Arthropoda being but one of them, in which structures show considerable diversities.

A similar situation was mentioned at the end of Chapter Five, dealing with Protozoa. The view was there expressed that the large and diversified phyla were not necessarily to be considered as having been derived from a single source; that they should be looked upon as a grade of organization, or as a broad and widespread stage of animal evolution.

The idea of a grade, which in some instances might well be described as a super-phylum, is more realistic than the alternative view of the independent, monophyletic phylum of classical zoology. Unfortunately our knowledge is nowhere complete enough to enable us to dogmatize as to the truth. For many years to come different zoologists will hold different opinions, determined by the importance they attach to each separate piece of evidence.

In any phylum, and even in any class, rich in species, there is always to be found a proportion of its members which stand apart from their fellows, seeming in some respects to be abnormal, or at least unexpected. The differences are mostly due to their showing an unusual combination of the various characteristics to be seen in the majority of the 'normal' members. Among the animals undoubtedly arthropodan there are three groups of this nature. They are the Onychophora, Tardigrada and Linguatulidae, and their best places in our schemes have always been hard to determine. One suggestion is as follows:

Phylum Pararthropoda
    Class Onychophora
    Class Tardigrada
    Class Linguatulidae
Phylum Arthropoda
  Sub-phylum Proarthropoda
    Infra-phylum Trilobita
  Sub-phylum Euarthropoda
    Infra-phylum Pycnogonida
    Infra-phylum Chelicerata
    Infra-phylum Mandibulata

This is an original and convenient arrangement. In this book the Onychophora have already been given separate treatment, and Trilobites, which are extinct, are mentioned below. The Euarthropoda ('true Arthropoda') divide themselves satisfactorily into a crustacean-insect-myriapod portion and an arachnid portion: the Crustacea and Insecta are more closely related to one another than either is to the Arachnida. The Pycnogonida bear a superficial resemblance to the Arachnida, with which class they have sometimes been associated. The chief reason for this is their possession of four pairs of legs, but they are distinct in many ways, which seems to justify their isolation in a group by themselves.

The remaining animals which may be thought of as ordinary or typical Arthropoda are naturally and obviously divided into those with mandibles and antennae, and those with chelicerae and pedipalpi. They therefore satisfactorily form the infraphyla Mandibulata (or Antennata) and the Chelicerata.

All the above considerations are clearly the result of the conspicuous success of the Arthropodan way of life. By 'success' a zoologist implies the existence of many species widely distributed, and occupying a variety of habitats, and a possible reason for this outstanding feature of the Arthropoda should be examined.

Their success is probably due to the elaboration of the cuticle and to the separation of the somites into separate parts with distinct contents and functions. An animal like an insect, with its mouth parts, eyes, and antennae on its head, its legs and wings on its thorax, and its viscera in its abdomen is built to a design that allows variation in any of these parts without any undue disturbance of the rest. The result must be a potentiality for adaptation to a much wider variety of environmental conditions than is to be found in the simpler Annelida.

The cuticle, however, is of greater significance, and is a much more complex coating than the thin and often transparent cuticle of many annelids. It is outside a layer of cells, known as the epidermis, which secrete it, and it consists of two distinguishable layers, a lower endocuticle and an upper epicuticle.

The endocuticle is based on chitin, a polymer of acetyl glycosamine, a resistant compound, insoluble in most reagents. In its lower layers it is often associated with calcium carbonate, usually in microcrystalline form, adding greatly to its mechanical strength. The upper portion, or exocuticle, is darkened ('tanned') by admixture with polyphenols.

The epicuticle is also laminated. Its lowest layer consists of a hard protein, cuticulin; over this there may be a layer of waxy or fatty material, which, though absent from some orders, is of great value since it is largely impermeable to water vapour. To terrestrial invertebrates this is of great importance, since it prevents or retards desiccation and dangerous concentration of the body fluids. Overall the epicuticle is covered with a layer of cement.

To these features there must be added the great capacity of the jointed appendage to assume almost countless varieties of form for all purposes and in all circumstances. When all these characteristics are combined with a fertility which results in incredibly rapid multiplication when circumstances are favourable, the universal biological success of the Arthropoda need cause no surprise.

Perhaps surprise is felt at the limitation of that success rather than its extent. Man wages an unceasing war against invertebrate pests, and often it seems that very little is needed to tip the scale seriously against him. What are the obstacles to the arthropods' progress?

The first is their size, which is limited by the existence of the exoskeleton. The weight of a large land arthropod would necessitate a thick and heavy skeleton, which is not consistent with periodic growth by ecdysis, for on emergence from its cold cuticle the heavy and soft animal might well be incapable of retaining its shape. Moreover, the new exoskeleton would take a proportionately longer time to harden, and so increase the danger period during which the animal could neither protect nor feed itself.

A second limitation is psychological. We are often lost in wonder at the marvels of instinct, and by instinctive behaviour an arthropod is able to survive. But instinct is not as good as intelligence and foresight, which Arthropoda exhibit rarely and to a limited extent. This may merely be because a small cerebral ganglion contains fewer neurones or nerve cells than a large brain, so that the invertebrate does not learn by experience or remember as effectively as a vertebrate.

Periodic growth quickly following ecdysis and ceasing as the new exoskeleton hardens is a characteristic of the phylum. Among insects it is known to be under the influence of hormones.

The neurosensory cells in the cerebral ganglia produce a fluid which stimulates the thoracic gland to secrete the hormone ecdyson. This causes the epidermal cells to construct a new exoskeleton under the existing one. The corpora allata, near the cerebral ganglia, produce another hormone, neotinin, which allows a larva to grow in size without changing into a pupa. When the larva has attained a certain size the secretion of neotinin ceases and the ecdyson now induces moulting or metamorphosis.

It follows that early removal of the corpora allata from a young larva results in the production of a small pupa and consequently of a small imago. Conversely, the implantation of corpora allata into a large larva will result in the formation of a giant larva, a giant pupa and a giant imago. These are not hypothetical consequences: they have been found to occur in the circumstances mentioned. Small quantities of the hormones have been extracted from large quantities of silkworm larvae, and have been found to be active in the tissues of other species.

No introduction to the Arthropoda should omit all mention of the fascinating extinct group known as the Trilobita.

These animals were evidently very abundant in the moderate depths of the Cambrian and Silurian seas, but they did not survive into the Triassic epoch. They were not large, most species being between one-and-a-half and three inches long, although a very few were almost ten times as long. Their bodies took the form of a flattened oval, the most conspicuous feature on the upper surface of which was a pair of longitudinal grooves, dividing each tergite into the three lobed regions from which the group takes its name.

The body consists of a uniformly covered head, evidently composed of five somites, a trunk of a varying number of relatively free somites, and a pygidium in which the tergites were usually fused and which ended in a telson. The pygidium could be flexed under the trunk, and the whole animal rolled into a ball like a woodlouse or a millipede.

The upper surface of the head carried sessile, compound eyes, the lower surface was a single labrum or hypostome below the mouth, with a small metastoma behind it. The only appendages of the head were a pair of uniramous segmented antennae, very probably homologous with the antennules of the Crustacea. All appendages of the trunk and pygidium were of the same pattern. They were biramous; the expodite was fringed with long bristle-like setae and bore at its proximal end a simple projection or gnathobase. It is possible that these fringes of setae may have been the means by which the animal collected its food; this is uncertain, but it recalls a similar arrangement found in some of the Branchiopoda today. The endopodite was a rod-like portion of five segments: there were no appendages on the terminal telson.

The trilobite egg hatched into a larva, known as a protaspis, which, almost circular in shape, consisted essentially of the head part of the adult. In this it recalls the nauplius, so widespread among Crustacea. There then appeared, attached to it, the ultimate somites of the pygidium, and later the middle somites of the trunk were inserted between them.

The interest in Trilobita has centred largely in their systematic position, or, the same thing, in their relation to modern Arthropoda. In every respect they are primitive animals. The larva, so like a protonauplius, the hypostome and metastoma guarding the mouth, the slender, jointed antennae, and the biramous limbs are all features in which they resemble the Crustacea. In 1902 Ray Lankester suggested that they were primitive Arachnida, from evidence derived from several fossils which seemed to link them with the Eurypterida. Their relation to any living arachnid, however, is not very close.

CHAPTER FIFTEEN

CRUSTACEA

Aquatic Arthropoda with a strong exoskeleton.
There are two pairs of antennae and three pairs
of manducatory appendages. Respiration is
effected by gills.

There are at least five sub-classes:

BRANCHIOPODA

Crustacea with immediately behind the head not less than
four pairs of flattened appendages, fringed with setae on their
inner edges. The eyes are compound.

OSTRACODA

Crustacea with a bivalved carapace and not more than two
pairs of trunk appendages which are rod-like, not flattened.

COPEPODA

Crustacea without eyes or carapace, and usually with six
pairs of thoracic and no abdominal appendages.

CIRRIPEDIA

Sessile Crustacea, often parasitic and normally hermaphrodite; the carapace covers the whole body. Typically there are
six pairs of thoracic appendages.

MALACOSTRACA

Crustacea, normally stalk-eyed, the carapace covering the
thorax of eight somites. The abdomen of six (sometimes seven)
somites, which bear appendages.

The Crustacea form a well-characterized and numerous class
whose members, with very few exceptions, have never forsaken
the water for the land. Largely for this reason they have not
attracted the same attention as the insects or even the arachnids. As marine and freshwater animals, however, they are
universally distributed, and the large proportion of small,
active species, which play a vital part in the economy of nature,
well justifies their popular description as the insects of the sea.

The obvious external feature of a crustacean is the hard carapace. In its typical form this covers the cephalothorax and, hanging down on each side, produces spaces into which the gills project while a current of water passes continuously over them. Some orders have no such carapace and a few have discarded it. It is often strengthened by the inclusion of crystalline calcium carbonate which may make it protective armour of a very resistant character.

The appendages or limbs of Crustacea show a wide range of variation. Two basic types are distinguishable, known as the

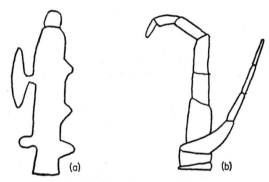

FIG. 23. Phyllopodium and stenopodium

flattened phyllopodia and the rounded stenopodia. The former type occurs chiefly in the sub-class Branchiopoda and is flexible and unjointed. It usually consists of an outer flabellum and epipodite and, on the inner side, of a row of endites, the basal one of which is the gnathobase (Fig. 23(a)). The stenopodium, sometimes called the biramous limb, consists of a basal protopodite, often carrying an epipodite and acting as the base or foundation of two branches, (rami), the expodite and endopodite. Usually the endopodite is the larger and more active branch, the expodite is homologous with the flabellum of the phyllopod limb (Fig. 23(b)).

Changes in the shapes and in the relative sizes of the parts of either type of limb can result in organs of very different appearances, adapted for several different functions, while the complete suppression of the expodite, as is not infrequent, may result in the formation of a uniramous leg-like limb. This great diversity is one of the chief features of the external parts of the Crustacea.

The body of a crustacean, at least in the higher orders, can usually be divided into head, thorax and abdomen, but fusion of the first two converts them into a cephalothorax.

The head is composed of five somites and its appendages are the antennules, antennae, mandibles, maxillulae and maxillae. It also carries the eyes, which are of two kinds, simple ocelli or compound eyes, often borne on stalks and with many surface facets, each of which acts as a lens.

Eight somites may in general be regarded as forming the thoracic region. The first three somites carry appendages of a type intermediate between the mouth parts and the walking legs and known as maxillipedes. They are followed by the chelae, which in lobsters and crabs may be greatly enlarged and serve as weapons of offence and defence. Behind the chelae are four pairs of legs which may or may not be chelate.

Six somites followed by the telson compose the abdomen. Its appendages are the pleopods or swimmerets, the anterior pairs of which may be modified to function as accessory sexual organs.

An examination of the numerous appendages, which is imposed upon most students of biology, shows that while they differ in functions they can all be traced back to one basic structural plan. The second maxilliped of a crayfish is customarily taken as representing the nearest approach to the original design, and the noting of its variations as shown in other appendages is one of the most fascinating occupations in the study of the higher Crustacea. Alternatively and equally attractive is the making of a simple comparison between the crayfish of the elementary laboratory and the prawn from a fishmonger's shop.

Such variations on a theme are to be found elsewhere in the animal kingdom. Among invertebrates a very admirable example is provided by the mouth parts of insects, and among the vertebrates by the almost equally diverse forms that are assumed by the pentadactyl limb. Such homologies and examples of adaptive radiation can only be reasonably interpreted in the light of the theory of evolution.

Internally, the Crustacea illustrate very well the nature of an invertebrate body in which the coelom is reduced and the body cavity is almost wholly haemocoelic. The alimentary canal is normally a straight tube, served by the gnathites in front and lined both anteriorly and posteriorly by invaginations of the cuticle. The absorptive area is thus limited to the mid-gut.

The heart occupies a mid-dorsal position just below the carapace and is normally enclosed in a pericardium. It contains oxygenated blood which travels in arteries to the tissues, collects in suitable sinuses for re-oxygenation in the gills, and finally enters the pericardium for return to the heart. The blood is almost colourless, for in place of red haemoglobin it contains haemocyanin, in which copper takes the place of iron. The gills are thin-walled outgrowths from the side of the body or from the proximal segments of the legs and combine the functions of osmo-regulation, oxygenation and sometimes filter-feeding, so that it is often necessary for the constant movement of one pair of appendages to direct an uninterrupted flow of water over them.

The nervous system is usually of the typical annelid-arthropod type, with a nerve collar round the pharynx uniting pairs of supra- and sub-pharyngeal ganglia, with a ventral nerve cord from the latter to the end of the body.

The sexes are separate, but many species can produce parthenogenetic eggs at certain seasons, and in other species no males have yet been described. The gonads are usually situated in the upper part of the thoracic region and owing to the absence of a coelom they are in direct connection with their gonaducts.

Generally the egg hatches into an ovoid creature known as a nauplius. This has one median eye, a pair of uniramous antennules, a pair of biramous antennae and a similar pair of mandibles. The last two have spine-like gnathobases at the sides of the mouth, and there is also a large labrum. A nauplius larva occurs in at least some groups of every crustacean sub-class, if perhaps only in the most primitive orders, or in a modified form. For example, the nauplius in the Ostracoda is covered by a carapace. Equally in every sub-class there are instances of its suppression, so that it is to be found only in the egg or in the brood pouch.

In some Crustacea the nauplius develops gradually, adding somites between the thoracic part and the telson while the appendages slowly assume their adult forms. In others the nauplius changes suddenly into an intermediate which precedes the adult: for example, the nauplius larva of the barnacles becomes a freely swimming cypris larva before it settles. In others again the first free larval form may be that remarkable object, the zoaea (Fig. 24). This is found among the crabs and other Malacostraca. It has a well-segmented abdomen and a cephalothorax fully provided with appendages in front but temporarily undeveloped behind.

Fig. 24. A zoaea larva

The first four sub-classes may be passed over briefly, but each has certain features of zoological interest and certain common species which call for mention.

BRANCHIOPODA

The two most attractive members of this group are Cheiro-cephalus diaphanus and Apus cancriformis. Both are fresh-water forms, occasionally found in Britain, and both swim upside down. The fairy shrimp, C. diaphanus, is about an inch long, the transparency of its body relieved by its red legs and telson. It is completely segmented, with eleven somites of the trunk bearing appendages. These are the flattened plate-like type already described as phyllopodia, and by their rhythmical beating the animal swims rapidly. It is the iridescence that follows their movements that gives the species its popular name.

The twelfth thoracic somite of the male carries a pair of processes, that of the female a median egg sac. The abdomen consists of seven somites without appendages and a telson which is characteristically forked.

Cheirocephalus belongs to the order Anostraca, Apus to the order Notostraca. The latter species is protected by a broad carapace above the trunk, and the abdominal somites carry from two to five pairs of appendages each. Males are rare.

Both these crustaceans and their relatives such as Branchippus (Fig. 25), can withstand the evaporation of the water in which they live. Their eggs remain viable in the dry mud, and if this is blown by the wind or carried on the feet of birds to water elsewhere they hatch out and resume life. The irregular

FIG. 25. Branchippus

appearance and disappearance of Apus in Britain is probably due to its sporadic immigration in this way.

A third species, belonging to the order Cladocera, is the familiar 'water-flea', Daphnia pulex, common in ponds and ditches in this country. Its carapace covers it like a bivalve shell and two branched antennae project from the head in front. The movements of the antennae cause Daphnia to swim in a characteristically jerky manner, hence the name water-flea. The chief function of the limbs is the collection of food particles, nearly all the Branchiopoda being filter-feeders.

OSTRACODA

This is a small sub-class of microscopic Crustacea, abounding in salt and fresh water. Many species are wholly parasitic.

COPEPODA

This is a very widely distributed group, in which the most familiar species is Cyclops fuscus (Fig. 26). Its oval trunk with its single eye, its forked tail and, in the female, the two oval egg sacs hanging from the body, make this an unmistakable crustacean. Many Copepoda, such as Calanus, occur in the

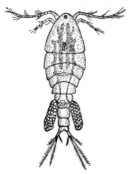

FIG. 26. Cyclops fuscus

marine plankton and are valuable food for herrings and mackerel; others, the 'fish-lice', are external parasites of other fish to whose skin they cling by means of their hooked claws while they suck their blood.

CIRRIPEDIA

This extremely interesting sub-class contains the barnacles and their allies, which have adopted a sedentary mode of life and in so doing have lost all outward appearance of Crustacea and have come to look superficially like Mollusca. Everyone who has been on a rocky shore has seen the little acorn barnacle, Balanus balanoides, often in very large numbers. Their bodies are covered with conical shells fixed directly to the rock, and in the centre there is an opening, guarded by two pairs of valves. When these are separated the feet of the barnacle protrude and keep up a constant movement.

The goose barnacle, Lepas anatifera (Fig. 27), is a much larger animal which is found in numbers on floating logs and is one of the chief causes of the fouling of the bottoms of ships. It consists of a muscular stalk two or more inches long, carrying at its free end a group of five calcareous plates. The stalk is derived from the head of the larva, and two minute antennae still exist in the cement which fixes the sessile adult to its base.

Six pairs of appendages project from the shell and by continually moving outwards, waving and retracting, bring food into the barnacle's mouth.

This crustacean has been famous since the Middle Ages because of the belief that it developed into the Barnacle Goose.

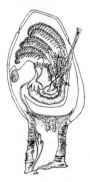

FIG. 27. Lepas anatifera, the goose barnacle

A member of this group which should not go unmentioned is Sacculina carcini, a parasite of crabs. The larvae settle on the crab's body and undergo a remarkable series of changes in the course of which fine threads grow into the crab. The parasite itself meanwhile degenerates into a featureless sac with no resemblance to any crustacean. The threads extract nourishment from every part of the crab, and by invading the testes produce parasitic castration of a unique kind.

MALACOSTRACA

This is much the largest and most varied sub-class of the Crustacea, yet despite its diversity all its members agree in showing six, eight and six somites to head, thorax and abdomen, with the female orifice on the sixth and the male orifice on the eighth thoracic somite. There are also several features of the appendages that occur only in this group. For example, the antennulae are biramous, the antennae possess a flattened leaf-like exopodite which acts as a stabilizer, the mandibles carry palpi, the thoracic appendages bear epipodites in the form of gills, and the abdominal appendages are biramous.

Five orders are recognized, but only the two largest, the

102

Peracarida and the Eucarida, will be mentioned. The clearest differences between the two are that in the Peracarida the carapace covers four, and in the Eucarida six thoracic somites. Internally, the former have a tubular, the latter a short heart.

PERACARIDA

The two most interesting groups of this order are the Amphipoda and the Isopoda. The former have a laterally compressed body, with no carapace and with sessile eyes. The freshwater flea, Gammarus pulex, is valuable as food for trout and other fishes, while on the beaches 'sandhoppers' are often disturbed in large numbers when heaps of seaweed are turned over. Many Amphipoda occur in the marine plankton: others, of the family Cyamidae, are the whale-lice. These live on the skins of whales, which they gnaw, and are peculiar in that they cannot swim and can only move from one host to another when the two are

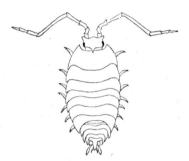

FIG. 28. Oniscus, a woodlouse

actually touching. In consequence one whale may carry several whole broods or even generations of Cyamus, amounting in all to some thousands of individuals.

The Isopoda or woodlice are the most unusual, or the most ambitious, of all Crustacea, since they alone are efficiently adapted to life on land. The appearance of the woodlouse is familiar to all (Fig. 28): an oval segmented body with the six somites of the abdomen narrower than the others. The head bears two small eyes and two pairs of antennae, the first pair small, the second long and active. The mouth parts (Fig. 29) consist of mandibles, two pairs of maxillae and one pair of maxillipeds, all of which are toothed for biting.

103

Like all animals with aquatic ancestors and relatives, wood-lice are constrained to live in a moist atmosphere. This is most

Fig. 29. Mandibles and first maxillae of Oniscus

obvious in the case of our large species Ligia oceanica, which is seldom far from the seashore.

A woodlouse lays its eggs directly into a brood pouch between the second and fifth thoracic somites, where they develop.

EUCARIDA

This order contains the Decapoda and the Euphausiacea, the chief interest in the latter being that they are the animals known as 'krill' and form the chief food of whales.

The Decapoda contain the most highly evolved Crustacea, and include the lobsters, prawns, crayfishes and crabs. Astacus, the freshwater crayfish, and Nephrops, the Norwegian lobster, are familiar to all students of elementary biology. The general uniting feature is the possession of five pairs of walking legs behind the maxillipeds, but in lobsters and crayfish the first of these have greatly enlarged chelae. This pair therefore act as weapons and not as means of locomotion.

The crabs of the sub-order Brachyura are the short-tailed decapods. The abdominal somites are thin and are folded under the broadened cephalothorax; and the animal usually runs sideways or, traditionally, crabwise.

CHAPTER SIXTEEN

INSECTA

Terrestrial Arthropoda usually endowed with
the power of flight. The body is divided into
three regions: head, thorax and abdomen. The
head of six somites bears one pair of antennae
and three pairs of mouth parts. The thorax has a
pair of legs on each of its three somites, and
wings are borne on the second and third somites.
The abdomen of eleven somites is without appendages in the adult. Generally the life history
includes metamorphosis.

There are two sub-classes:

APTERYGOTA Insects without wings.

PTERYGOTA Insects with wings.

There are about fifty orders of insects (Fig. 30) and their
possible methods of grouping are given in Chapter Three. They
depend, to varying degrees, on the mode of development of the

FIG. 30. A generalized insect

wings, on the extent of metamorphosis during growth and on
the evidence of palaeontolgy.

The class Insecta, originally called Hexapoda, is unap-
proachable for range, variety of species, number of individuals
and economic importance. At least six hundred thousand

species have been described and, of all known species of animals, insects constitute about 80 per cent. Morphologically their chief interest lies in their wings and mouth parts, biologically in their life histories, and economically in their destructive and parasitic habits.

The head of an insect, composed of six united somites, carries the eyes and the antennae above and the mouth parts below. The eyes, like those of Crustacea, are either simple ocelli with a smooth lens formed from a transparent portion of the cuticle, or compound eyes with a many-faceted surface.

The antennae are uniramous—many-jointed organs of considerable versatility. They invariably carry a number of sensory setae which respond to touch; there are besides organs of chemo-reception which, reacting to chemical compounds in solution or as vapour, provide the same sort of information about the environment that human beings describe as tastes or smells. And in some insects there are organs which react to the humidity of the surrounding air.

The mouth parts are fewer than those of the Crustacea, there being typically only three pairs, the mandibles, the maxillae and the labium. Mandibles and maxillae are paired; the labium, which corresponds to the second maxillae, has become a median appendage, a lower lip formed by the fusion of the two halves in the middle line. In addition to these there is found a downward extension of the surface of the clypeus or forehead, called the labrum, and an extension of the dorsal surface of the pharynx called the epipharynx, and a similar extension of the ventral surface, called the hypopharynx. All these are as important in the life of the insect as in the work of the systematic entomologist.

This is largely because the mouth parts can be so widely modified in accordance with the nature of the animal's food and its method of obtaining it. Thus there are to be distinguished biting, piercing and sucking arrangements. The adaptations consist of changes in size and shape of the various parts and sometimes their loss. Some examples given below will make this clearer.

Biting mouth parts (Fig. 31) are considered to be the simplest and are found in cockroaches, earwigs, locusts and beetles. In these insects the mandibles are larger than in any others; they are hard and bear a number of sharp teeth. The value of such mandibles to an omnivorous feeder like a blackbeetle is obvious. The maxillae are among the most typical of all arthropodan manducatory appendages, with a jointed palp as exopodite and an endopodite divided into two blades, the lacinia and the galea. The labium is a median organ, formed by the fusion of the proximal segments of the paired second

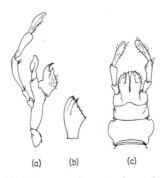

(a)　　　(b)　　　(c)

FIG. 31. Biting mouth parts of a cockroach
(a) Mandible. (b) Maxilla. (c) Labium

maxillae. It also carries palps and acts, as its name implies, rather like a lower lip.

Sucking mouth parts (Fig. 32) are familiarly seen in butterflies and moths. Since the food of these insects is generally nectar and needs neither breaking nor picking up, the mandibles

FIG. 32. Sucking mouth parts of a butterfly

are either absent or vestigial, and the maxillae have no palpi or very small ones. The lacinia is absent and the galea is converted into a long semi-cylinder, so that the two, placed together, form a tube. This is usually coiled under the head and can be straightened for penetration to the nectaries of flowers.

Ants, bees and wasps have mouth parts which can bite and

either suck or lick (Fig. 33); their action can best be understood by watching a wasp attacking a ripe plum. The mandibles, though small, resemble those of a cockroach but without the same strength. The maxillae are blade-like and are useful, for

FIG. 33. Mouth parts of a bee

example in shaping the masses of pollen and nectar which provide some of the food of the hive. They have small palps, while the palpi of the labium are much longer and their median part forms the proboscis.

FIG. 34. Piercing mouth parts of a mosquito

Piercing mouth parts (Fig. 34), are the most highly specialized and those of a mosquito provide a good example. Five rigid pointed lancets are derived from the mandibles, maxillae and hypopharynx. These lie in a trough-like labium. When a mosquito 'bites' the lancets penetrate the skin with surprising ease, while the labium bends behind them and salivary glands pour a nanti-coagulant into the wound, mixed perhaps with malarial or other parasites.

Organs which also pierce and suck are found in the Hemiptera. The labium surrounds four stylets, roofed at their base by the epipharynx and derived from mandibles and maxillae. They are grooved on their inner faces so that when in contact they form channels up which the fluids of the animal or plant victims can be imbibed. A very similar arrangement is found in the Anoplura or sucking lice.

APTERYGOTA

COLLEMBOLA

Springtails are widely distributed insects, found both in fresh water and on land, sometimes in immense numbers and extending beyond both the arctic and the antarctic circles. The lifting of a heap of drifted leaves often disturbs them, leaping vigorously in their characteristic fashion. A common British species, easily recognized by the golden band across its back, is Orchesella cincta.

The head carries biting mouth parts and the thorax behind it has three pairs of legs which have no tarsi, their claws being attached to the tibiae. The abdomen is composed of six somites only. From the first of these there projects a ventral tube which seems to secrete an adhesive substance. The third bears a catch which engages the fork on the fourth somite. The leap is the consequence of the release of the fork by the catch. The first insects discovered in the Antarctic, Gomphiocephalus hodgsoni, belong to this group, which in the ways mentioned above is so different from all other insects that some entomologists exclude them from the class.

PROTURA

This is perhaps the most primitive of all the undeniable insects. The common British species is Campodea which has no eyes, twelve somites to its abdomen, and a pair of long anal cerci.

THYSANURA

The silverfish, Lepisma saccharina (Fig. 35), the firebrat, Thermobia furnorum, and the seashore Machilis maritimus, are members of this order. The silverfish is very common in and near houses, where their little silvery bodies are often seen running in the evening among household stores. They do not eat enough to be a serious nuisance, except when they sometimes turn their attention to paper and gnaw the edges of books.

Like the members of the last two orders, they have no wings and develop without any metamorphosis. The maxillae of their

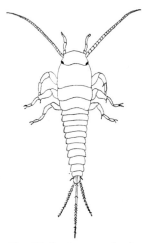

Fig. 35. Lepisma saccharina

mouth parts are remarkably similar to those of Crustacea and they are accompanied by a pair of so-called maxillules. This is another crustacean feature and is found in no other insect order.

The abdomen is of ten somites, some of which carry small rod-like organs at the hind edge of the sternites. Their function is obscure, as is that of several small vesicles placed near them and capable of being withdrawn into the body or extruded from it. The tenth somite carries three long, jointed cerci.

The female lays from seven to twelve eggs, dropped haphazard. The firebrat is rather larger, darker in colour and slightly hairy. It lives only in warmer places. The semi-marine Machilis has black scales among the silvery ones in something of a pattern and its middle cercus is longer than the others.

The orders of insects are too many for all to be treated in this book, and the most reasonable course to take is to offer some description of the essential characters of a selection.

EPHEMEROPTERA

The mayfly's eggs are laid in water and hatch into nymphs which in their later instars have pairs of tracheae-containing gills on the abdomen. They have biting mouth parts and are vegetable feeders. Their nymphal life is very long and comprises over twenty instars before a winged sub-imago appears. This flies until, within the period of a day, it has moulted for the last time to become mature. The adult has rudimentary mouth parts which do not permit it to feed and the alimentary canal is full of air. The adults mate and die very soon after oviposition. Both nymphs and imagos are freely eaten by fish.

ODONATA

The dragonfly (Fig. 36) lays its eggs in water or on plants growing in water and the nymphs are aquatic animals breathing by tracheal gills. In some species these gills are on the side

Fig. 36. A dragonfly

of the rectum and water is continuously sucked in and driven out through the anus. Rapid expulsion of the water helps the nymph to dart forward.

This immature form of the dragonfly has an enlarged labium known as the mask, which it can suddenly shoot out, piercing its prey with the labial hooks. The mature form is a large insect and is a powerful and rapid flyer which catches its victims in the air, using its forelegs for this purpose. It has four large wings equal in size, prominent eyes and small antennae. Dragonflies mate while flying.

DICTYOPTERA

Cockroaches are insects that are numerous and easily obtained, large enough for beginners to dissect successfully, and sufficiently unspecialized to provide a good picture of insect anatomy. They are consequently familiar to every zoologist (Fig. 37).

Their anterior wings are stout and firm, and the posterior

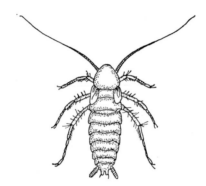

Fig. 37. Periplaneta americana, black beetle

wings, which are membranous, are folded beneath them. They have typical, simple, biting mouth parts.

Characteristic features of the domestic black beetle, Periplaneta americana, are the forward extension of the first thoracic tergite, under which the head is hidden, the length of the coxae, which gives the insect its remarkable speed, and the pair of cerci anales projecting from the end of the abdomen. In the male there is also a pair of styles.

A pair of large salivary glands, each consisting of two groups of secretory cells and a reservoir, occupy much of the interior of the thorax. Cockroaches are omnivorous, and in temperate countries are almost confined to houses, but many other species of the order live normally out of doors. The female lays sixteen eggs at a time, enclosed in an oötheca produced by special accessory glands.

The praying mantis, Mantis religiosa, belongs to this order. The femora have a ventral groove with strong spines on its edges, the sharp tibia fits into it, and the two segments make an efficient means of seizing and holding the prey.

ISOPTERA

The white ants have evolved an elaborate social organization which surpasses that of any other animal, not excluding man. A colony of termites occupies a nest or termitary which may be a maze of tunnels in a tree or in the ground, or may be a structure built of earth cemented with saliva. The inhabitants consist of:

i The royal pair, or king and queen, whose wings have been removed
ii winged fertile individuals for new colonies
iii wingless fertile individuals for emergencies
iv sterile workers for labour
v sterile soldiers for battle
vi nymphs

The king and queen after mating dig a small burrow and the first offspring are tended by their parents until they can help with the up-bringing of their younger brothers and sisters. The queen's wings are broken off and she rapidly grows to an enormous size, becoming so inert that she must be fed by the workers and can devote the whole of her energies to the laying of eggs. This she does with such success that a total of thirty thousand eggs a day has been recorded, and her total output may approach a million.

The sterile workers, which are none the less potential males and females, have larger mandibles than the nymphs. They are the insects that do such damage to human property, consuming furniture, books and other objects at an incredible speed, and they are able to digest cellulose because of the presence of symbiotic Protozoa in their alimentary tract.

ORTHOPTERA

Grasshoppers and crickets are known to everyone, and locusts have for long been one of the most dreaded pests to which mankind may be exposed. These are all large insects whose most obvious characteristic is the strength of the femora of the third pair of legs, enabling the animals to leap very considerable distances. They can also make peculiar noises, known as stridulation. In the larger grasshoppers (Tettigoniidae) the sound is produced by rubbing together the tegmina or first wings; the domestic cricket, Gryllus domesticus, uses the same method: the smaller grasshoppers and the locusts rub the rough inner surface of the femora against the tegmina.

Locusts (Acridiidae) are gregarious, and in certain favourable conditions multiply fantastically, producing the migratory phase. These then take to the wing in vast swarms of tens of millions, and the ultimate settling of the multitude in cultivated districts results in utter desolation of every edible crop. Between migrations the locust lives a solitary and harmless life in a form markedly different from the migrants.

DERMAPTERA

Earwigs form a small order of nocturnal insects, omnivorous but seldom obnoxious, and credited with the unusual reputation of being good mothers. The eggs are laid in the soil, guarded from predators by the female, who also cares for the nymphs. The posterior wings when unfolded are almost circular

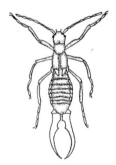

FIG. 38. Forficula auricularia, male earwig

in shape, but normally are folded both transversely and longitudinally under the elytra, and have very seldom been seen in use by the common British earwig, Forficula auricularia (Fig. 38). Earwigs are always recognizable by the pair of forceps at the end of the abdomen, organs which are strong and curved in the male but straight and slight in the female.

COLEOPTERA

Beetles are characterized by the hardening of the fore-wings into thick elytra which meet in the middle line; the hind wings, folded beneath them, are membranous and are often reduced or absent. This is by far the largest insect order, with almost a quarter of a million described species, all of which have biting mouth parts, a large prothorax and a reduced mesothorax.

They are often serious pests to man, but zoologically their most interesting feature is the remarkable number of modifications in mode of life adopted by various species. Beetles may be found in almost every conceivable situation, living on almost every possible kind of food, in every part of the world. This is a phenomenon known as adaptive radiation and is better illustrated by Coleoptera than by any other order of animals.

Thus it is scarcely surprising to learn that about a hundred families of beetles have been described, and are divided among three sub-orders, Adephaga, Archostemata and Polyphaga, based on the patterns of the veins of their wings.

TRICHOPTERA

Caddis flies form an order of small importance and their chief interest is centred in their larvae. The adults congregate in mating swarms, after which the females lay their eggs in or near the water in which the larvae will live. Generally speaking, the larvae are of two kinds, slow-moving vegetable feeders and active predators. The former build round their bodies the cases for which the caddis flies are famous, closely-fitting protective sheaths made of small stones, shells or pieces of wood secured by a silk foundation. The predators spin webs of silk which trap small creatures washed into them by the stream; the only instance of a food-catching web outside the order of spiders.

The pupae are formed inside a silk cocoon which remains in the case. A current of water runs continuously through the pupal cocoon, and in due time the pupa bites its way out with its strong mandibles, makes its way to the bank, moults and emerges as an adult insect.

LEPIDOPTERA

Butterflies (Fig. 39) and moths have wings that are covered with scales by which their elaborate and often attractively coloured patterns are produced. Their diet is the nectar of flowers, hence their mouth parts, already mentioned, consist chiefly of a sucking proboscis, normally carried curled beneath the head.

Their two wings on each side are usually joined by a suitable system of hooks, so that in flight they rise and fall as one. Their designs cause some species to bear a close resemblance to other species. This coincidence is called mimicry, and the two insects are distinguished as model and mimic. The value of mimicry depends on the unpleasant taste of the model,

FIG. 39. A butterfly

which causes birds to refrain from eating it, and, by implication, to mistake the probably palatable mimic for the nauseous model they have learnt to avoid.

The sexes are separate and are distinguishable by appearance in only a few species. The male is attracted to the female by scent, which it is able to detect at great distances.

The eggs hatch as larvae, familiarly known as caterpillars and of interest because of their fully segmented form. Nearly all lepidopterous larvae are vegetable feeders, often to our great disadvantage, and their heads are furnished with a pair of toothed mandibles, together with maxillae and labium. Behind the head there are three thoracic somites carrying legs, and ten abdominal somites with prolegs on somites 3, 6 and 10. There are nine pairs of spiracles for respiration, and silk may be secreted from glands which open inside the mouth.

After its last ecdysis the larva changes into a pupa, which

may be unprotected or may be encased in a cocoon which the larva has produced. It is a passive, resting stage, during which the internal organs disintegrate and a new set of adult organs is produced from the white, creamy fluid which fills the pupa.

Lepidoptera are justifiably regarded as being among the most beautiful of all invertebrates, but in contrast to this, the Platyhedra gossypiella, the cotton boll weevil, and Porthetria dispar, the gypsy moth, are two of the most destructive pests in the world.

DIPTERA

Flies, in the conventional sense of the word, constitute a very large and highly specialized order of insects. Their most obvious diagnostic feature is their possession of the first pair of wings only, the posterior pair being converted into short knobbed projections, known as halteres and functioning as stabilizers. Correlated with this is the enlargement of the meso-thorax and the reduction in size of the other two thoracic somites.

Their mouth parts are primarily adapted for sucking and in many well-known species, such as gad-flies and female mosquitoes, are also designed for piercing the skin of the victim whose blood is to provide the meal. In this species the maxillae and mandibles are pointed stylets lying in a trough made from the labium, in which they are accompanied by a tube made from the opposed epipharynx and hypopharynx. The mandibles pierce the skin as easily as a hypodermic needle, the maxillae enlarge the wound, and the tube is thrust into it. Saliva pours down a canal in the hypopharynx and the host's blood is sucked in. In a similar way many Diptera suck up drops of fluid, often after their saliva has had time to dissolve the original material.

In the common blow-fly, mandibles and maxillae are missing and the labium holds only the epipharynx and hypopharynx. In this and other species, the labium ends in a pair of flat lobes marked with canals (pseudotracheae) along which the food passes. The whole proboscis can be folded under the head, but in Glossina and the stable-fly, Stomoxys, the pharynx is too rigid for this.

Diptera which do not suck blood are normally drinkers of plant juices, and it need scarcely be recalled that those which do feed on blood have become the vectors of the organisms responsible for malaria, elephantiasis, yellow fever and sleeping

117

sickness. To this may be added the fact that the adhesive pulvillus, the sticky pad on the tarsus, helps to carry pathogenic bacteria and Protozoa from all kinds of refuse to the food of human beings.

The most formidable of the Diptera are without doubt the gnats and mosquitoes, especially those that act as vectors for the parasites causing such diseases as malaria and yellow fever. These were mentioned in Chapter Five. The discovery of the cause of malaria by Sir Ronald Ross, and his demonstration of the part played by the mosquito Anopheles, was one of the greatest triumphs of zoology in the service of man.

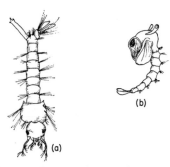

Fig. 40. Larva and pupa of a mosquito

From an entomological point of view mosquitoes are of interest because they lay their eggs on water and the larvae and pupae (Fig. 40) are aquatic. They are often to be seen on water butts when these are unprotected. They depend, however, on the air above for respiration and hence are readily controlled by spreading oil on the water surface. The change in surface tension makes it impossible for the animal to support itself from the surface film and so it suffocates or drowns.

The typical dipterous larva is a legless and practically headless 'grub', familiar to fishermen as gentles. The pupa is usually enclosed in the last larval skin inside a protective puparium. When the insect within is ready to emerge, this case is burst open by the inflation of a sac, known as a ptilinum, on the inhabitant's head, a unique and remarkable arrangement.

Even a minimal account of the Diptera must include at least a mention of some of its most familiar and notorious genera. Among these are Musca and Fannia, the larger and smaller house-flies, Calliophora erythrocephala, the blue-bottle, Lucilia caesar, the startlingly beautiful green-bottle, Tabanus, the gad-flies, Anopheles, Aedes and Theobaldia, mosquitoes, Glossina palpalis, the tsetse fly, Tipula the crane fly, Hypoderma the warble fly, not forgetting Drosophila, the most famous and most continuously bred of all laboratory flies; and it becomes clear that the ancient writers were justified in naming Baalzebub as the god of flies.

SIPHONAPTERA

Fleas (Fig. 41) are as well adapted to their mode of life as the Diptera are to theirs. As external parasites of birds and mammals they are without wings and depend for escape from danger on their phenomenal powers of jumping. Their bodies are flattened laterally and their tarsi end in two curved claws.

FIG. 41. Pulex irritans, a flea

Their mouth parts recall those of the Diptera. The palpi of the labium support the other appendages, the sharp mandibles pierce the skin and the blood rises up a channel formed by the mandibles and epipharynx.

Fleas lay their eggs among the feathers or fur of their hosts, but as they are not fixed they soon fall off and hatch on the ground. The larva, of a head and thirteen somites, is superficially like a dipterous larva: it undergoes three ecdyses, changes into a pupa and completes its life cycle in about a month.

In general each species of flea is limited to one host, and there is no appreciable chance of a human being suffering from the fleas of his dog, cat or chickens. The human flea, Pulex irritans, is a serious menace only when infestation is heavy, but a truly dangerous flea is Xenopsylla cheopis, the rat flea. This flea harbours Bacillus pestis in its gut, evacuates as it bites, and any scratching that follows may allow the bacteria to enter the host's body. Rats are thus responsible for the spread of plague.

HYMENOPTERA

Ants (Fig. 42), bees and wasps are insects which, like the termites, have evolved a social organization in which duties undertaken on behalf of the community have complete priority over individual needs, and the power of sexual reproduction is confined to a chosen few.

These are insects which typically have four membranous wings, the posterior pair smaller and joined to the anterior by

FIG. 42. An ant

a series of hooks. These engage with a groove on the front edge of the smaller wings. The mouth parts are basically of the biting type, but the maxillae and labium are united so that they act together, licking up food in the manner of a tongue. A wasp, watched feeding on a piece of over-ripe fruit, makes the action clearer than many lines of print. In some genera, such as Apis, the honey bee, the mouth parts reach a high degree of specialization. The mandibles are large, smooth spatulas, used for manipulating nectar or pollen. The labium carries the tongue, which projects forward with the concave labial palps at its sides, and with the paraglossae so encircling its base as to direct the food from the groove in the glossae to the mouth above it.

The social life, for which the order is so famous, seems to be in part the result of gradually changing food with which the larvae are supplied. Among the solitary wasps the egg is laid in or near sufficient food, e.g. a paralysed spider, to keep it alive until metamorphosis of the larva; the bee larva, on the other hand, is the object of progressive dietary. The duty of the workers follows an established time table, arranged so that the youngest are employed in domestic work in the hive, while the older ones either guard the entrance or fly out to collect pollen and nectar until, literally, they have worked themselves to death.

The order also contains the insects known as ichneumons and sawflies. The latter, placed in the sub-order Phytophaga, are the most primitive of the Hymenoptera. The ovipositor is really a saw, of two toothed plates, and is used to perforate the tissues of the plants in which the eggs are laid.

The ichneumon flies, or Parasitica, are a group of some four thousand species. Their habit of laying their eggs in the bodies of other insects has made them one of man's most valuable insect allies, controlling pests and maintaining the balance of nature.

PSOCOPTERA

The 'book-lice' are small insects not at all uncommon in houses in this country, where they are quite harmless. Some species are winged, and their wings are covered with scales very similar to the scales of butterflies. Other species are wingless and may be found living in rather dry air such as exists among books and stored manuscripts. Their larvae have been credited with making the tapping noise, vaguely attributed to the 'death-watch beetle', and supposed by the superstitious to be the heralding of bereavement.

ANOPLURA

Lice (Fig. 43), which are among the less generally attractive insects, have nevertheless many points of zoological interest.

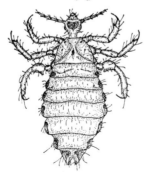

FIG. 43. Pediculus humanus, a louse

There are two sub-orders, the Mallophaga or biting lice, whose victims are usually birds, and the Siphunculata or sucking lice, which are parasites of mammals, including man. The common pest of man, Pediculus humanus, exists in two sub-species found on the head and body, with only minor differences. Lice lay their eggs attached to hairs or fibres of clothing at a rate of about ten eggs a day. Their life history, which includes three ecdyses, may be completed in three weeks. Their importance lies in their being the bearers of the pathogenic organisms known as Rickettsia, which escape from the louse's gut when it feeds. The scratching which often follows the irritation of the bite may allow the Rickettsia to enter the blood of the man. Here they are responsible for the diseases of typhus and relapsing fever, as well as for a modified form known during the First World War as trench fever. Today both lice and typhus may be controlled by the use of DDT.

An evolutionary relationship exists between the lice and their hosts. Allied species of birds are infested by allied species of lice, and the same relation exists between the lice of mammals and the species of mammals on which they are found. It appears that the evolution of lice has proceeded step by step with the evolution of their vertebrate hosts, an evolutionary pattern which is of great significance to phylogenists.

HEMIPTERS

The bugs form a huge order of nearly forty thousand species, showing much diversity within the group which includes water-boatmen, shield-bugs, pond-skaters, water-scorpion,

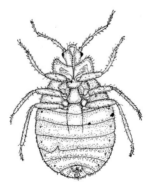

FIG. 44. Cimex lectularius, a bug

cicadas, plant-lice, scale-insects and aphids. The bed bug, Cimex lectularius (Fig. 44), is also a member of the order.

The most interesting zoological feature of the group is the fact that they alone among the Endopterygota possess piercing mouth parts. Mandibles and maxillae form four stylets, able to pierce plant tissues as well as animal skin, lying in a protective trough formed from the labium. This condition so closely resembles that found among the mosquitoes that it is an admirable example of the phenomenon of parallel evolution. Many bugs, of which the familiar Aphids provide an example, have evolved the power of virgin reproduction or parthenogenesis. Generations are produced with no males, yet without fertilization, the females produce living young directly or sometimes eggs that hatch normally. At intervals these parthenogenetic generations are replaced by an alternating generation including both males and females and producing viable eggs by fertilization.

CHAPTER SEVENTEEN

CHILOPODA

Elongate Arthropoda of many distinct somites, each of which bears one pair of legs. The body is somewhat flattened dorsoventrally. The head consists of four somites and carries the antennae, mandibles, maxillae and palpi. The first somite of the body proper carries venom-producing maxillipeds; the penultimate somite has a specialized pair of claspers; the genital orifice is on the last somite. Respiration is by tracheae, which open at spiracles in the pleural membrane.

There are five orders:

GEOPHILOMORPHA

Chilopoda of from 31 to 177 pairs of legs, antennae of 14 segments, and with spiracles on all somites except the first and last.

SCOLOPENROMORPHA

Chilopoda with not more than 23 pairs of legs, and with antennae of 17 to 30 segments. Spiracles are not present on every somite.

CRATEROSTIGMOMORPHA

Australasian Chilopoda with 15 pairs of legs and 21 tergites.

LITHOBIOMORPHA

Chilopoda with 15 pairs of legs and antennae of 20 to 50 segments. There are 6 or 7 pairs of spiracles.

SCUTIGEROMORPHA

Chilopoda with 15 pairs of very long legs, and with 8 tergites and 15 sternites.

The first two of these orders are sometimes put together in a sub-class Epimorpha and the others in a sub-class Anamorpha.

Centipedes (Fig. 45) are of zoological interest because they are familiar to everybody and provide an example of an arthropod as fully segmented as an annelid. If one of the polychaete Annelida became adapted to life on land it would probably be very like a centipede.

A centipede's body is divisible into head and trunk; and, since the animal is carnivorous, the head, which is a flattened

Fig. 45. Lithobius, a centipede

lens-shape, carries a set of mouth parts as well as a pair of antennae. There are also groups of simple ocelli.

The antennae are flexible chemotactic organs of many segments whose number is used by systematists to help characterize the different orders.

The mouth parts (Fig. 46) are the mandibles and two pairs of maxillae. The mandibles, one on each side of the pharynx, are toothed organs designed for biting. The first maxillae are fused together in the middle line to form a median plate composed of two coxae and their sternites, and therefore known as the coxosternite. At each side is a short maxillary palp. The second maxillae lie closely behind the first and are similarly composed of a coxosternite and a pair of palpi. The function of these maxillae is to act as a lower lip and direct food particles towards the mandibles for mastication.

In the trunk the first and last somites are different from the rest. The basilar or forcipular somite, immediately behind the head, carries a pair of maxillipeds of four segments, the last being a pointed fang perforated by the duct of the poison gland. The proximal segments, like those of the maxillae, are

125

fused and also act as a lower lip. There is a recognizable resemblance between the mouth parts of a centipede and those of either a crustacean or a primitive insect: the chief difference

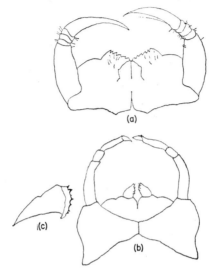

Fig. 46. Mouth parts of Lithobius
(a) Maxillipeds. (b) First maxillae. (c) Mandible

is that in a centipede the maxillae are relatively smaller and the maxillipeds larger.

All succeeding somites except the last two carry a pair of legs of seven segments and ending in claws. On the penultimate somite is the orifice of the reproductive system, lying between a pair of modified appendages known as gonopods (Fig. 47). The last somite is smaller than all the others and is perforated by the anus.

Fig. 47. Terminal somites of Lithobius

126

The internal organs of a centipede are not complicated. The alimentary canal has a short pharynx into which open two or sometimes three pairs of salivary glands. The mid-gut is a long straight tube with no adjoining glands, leading to the rectum at the beginning of which is a group of Malpighian tubules.

The vascular system is well developed, with a heart running the whole length of the body. In each somite are two ostia and a pair of lateral blood vessels. Forward from the heart runs the cephalic artery as well as a pair of vessels that surround the pharynx and then unite to form the supraneural artery.

Respiration is by means of tracheae as in insects and arachnids.

The reproductive system of the female consist of a median ovary above the alimentary tract. Two oviducts lead from it to a wide genital pouch which is in communication with two spermathecae and accessory glands. The male has from seven to thirteen pair of testes united to a long coiled epididymis and vas deferens.

The most interesting glands in centipedes are the venom glands in the maxillipeds, with ducts which open at the tips. The acid venom is quickly fatal to the centipede's normal prey, such as worms, spiders and insects. It is really serious to man, and though only rarely fatal, yet there is pronounced fear of the centipede in many countries.

Centipedes nearly always live in damp, dark places; they are common in caves and a few are found near the sea. They tend to seek a moisture-laden atmosphere because there is no wax layer in the epidermis, such as prevents or delays the loss of water from the bodies of many arthropods.

A female centipede lays from fifteen to twenty eggs loosely in the ground. The young of some species are born with the full number of legs.

Commonly grouped with the Chilopoda in the now abandoned class of Myriapoda were the orders Pauropoda and Symphyla.

The Pauropoda are small inhabitants of the damp spots under leaves and logs and in the first few inches of soil. They seldom exceed 2 mm in length, and their bodies consist of a head and twelve segments with ten pairs of legs. As in millipedes, the genital orifice is near the front end of the body. Five pairs of sensory setae project from each side of the animal.

Pauropoda are typical members of the cryptozoic fauna, and

are probably fungus feeders. The eggs hatch as larvae with five or fewer pairs of legs and increase the number as they grow by moulting.

The Symphyla are rather larger, and may approach half a centimetre in length. Their antennae are long and there are twelve pairs of legs and fifteen somites. Spinning glands open at a pair of processes on the last segment. These animals live like the Pauropoda in damp litter, but they are better-known. The species Scutigerella immaculata occurs in Europe, Africa and America; other species though less well-known are cosmopolitan. They are rapid movers and are apparently vegetarians. Some zoologists believe that they are related to the ancestors of the insects.

CHAPTER EIGHTEEN

DIPLOPODA

Elongate Arthropoda with two pairs of legs on most somites. There is a distinct head, carrying a pair of antennae and at least two pairs of mouth parts. The body is rounded in cross-section, and the tergites are larger than the corresponding sternites. Spiracles leading to respiratory tracheae are found above the coxae of the legs. The genital orifice opens on the ventral surface of one of the anterior somites. There are odoriferous glands in many genera, but no venom glands and the animals are generally vegetarian.

There are two sub-classes:

CHILOGNATHA

Diplopoda in which the exoskeleton is hard and the head carries two pairs of gnathites.

PSELAPHOGNATHA

Diplopoda in which the exoskeleton is soft and the head carries four pairs of gnathites.

The head, like that of the Chilopoda, consists of four somites. It carries the seven-jointed antennae, the mandibles, the first maxillae, which are present only in the embryo and have vanished from the adult, and a pair of persistent second maxillae.

The mandibles are of two or three segments and have no palpi: the maxillae, formed by fusion of paired appendages, are known as the gnathochilarium. This is in close association with the collum, the somite that follows the head and which has no appendages and no stigmata. It will be observed that there are no venom glands associated with these very characteristic mouth parts, for the Diplopoda are purely vegetarians.

A further characteristic feature of the diplopod body (Fig. 48) appears in the next three somites. These differ from the rest in the possession of a single pair of legs each, so that to this extent a thorax, or at least a thoracic region, may be distinguished from an abdominal, posterior region.

In Diplopoda the genital aperture is situated between the second pair of legs and not on the last somite.

In the posterior part of the body there are apparently two pairs of legs to each somite, the result of an early fusion of the somites in couples. The somites have tergites that are longer

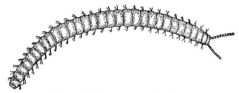

Fig. 48. Julus, a millipede

than the sternites and are narrower in front than behind, an arrangement that facilitates the animal's habit of rolling itself into a ball. In general the somites are less flattened than are those of the Chilopoda, so that a section of a body of a millipede is round rather than oval.

Internally, the alimentary canal is a straight tube, associated with a large number of glands in relation to the vegetable diet of a millipede.

A conspicuous feature of a millipede's body is the toughness of its cuticle, which is hardened like that of an arachnid and contains calcium salts like that of a crustacean. Added protection is afforded by their ability to secrete, slowly, or sometimes in jets of some force, a repellant fluid which is dangerous to human eyes.

Although they have no eyes, millipedes usually avoid the light, and like centipedes they are constrained to live in moist environments. Normally their food consists of the vegetable material found in humus, but in some circumstances they may be impelled to migrate, occasionally in large numbers, when they may do serious damage to crops.

Fertilization is internal and the eggs are laid sometimes in the ground and sometimes in carefully constructed nests where they receive a degree of maternal protection.

Clearly the differences between Chilopoda and Diplopoda are considerable, and justify their separation into two classes.

It is, however, of interest to note that all four groups of the old Myriapoda show a common and unusual feature, sometimes described as diplopodism. This is the formation of doubled somites, which is not confined to the true Diplopoda. It is found in all myriapods, for in all four classes there are some genera in which the number of visible tergites is smaller than the number of sternites. It is a peculiar phenomenon, not easily understood.

CHAPTER NINETEEN

XIPHOSURA

Aquatic Chelicerata in which the prosoma is horseshoe-shaped and bears simple median and compound lateral eyes. The chelicerae are of three segments and are chelate: there are five other pairs of leg-like appendages. The opisthosoma is of ten somites and carries plate-like appendages. There is a telson in the form of a spine.

This group deserves mention because it contains the famous king crab, Limulus polyphemus (Fig. 49), which, with the allied genera Tachypleus and Carcinoscorpius, represent the survivors of the ancestral Chelicerata, the class Merostomata.

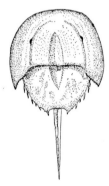

FIG. 49. Limulus polyphemus, the king crab

King crabs are found in comparatively shallow water on the east coast of North America, and also in Malaysia, China, Japan and India. The almost semicircular prosoma has sloping sides, so that the carapace forms an inverted bowl in which the appendages are hidden from sight. The chelicerae, three-jointed chelate organs, are unremarkable; a more noteworthy feature is that the five pairs of limbs that follow are very similar and the foremost of them cannot reasonably be distinguished as pedipalpi. The legs of the last pair are rather different since

132

they carry outgrowths which open under pressure to form a scoop with which the king crab pushes away the sand as it buries itself.

An unexpected feature of the class is the persistence of appendages in the middle region of the body or mesosoma. The tergites of the six segments involved are fused together and form a part of the general carapace, but underneath six appendages are found. They take the unusual form of flattened plates. The first pair are the largest: known as the genital operculum, they cover the simple opening of the genital system. The other five carry gill-books or respiratory organs on their forward edges, an arrangement that is vaguely reminiscent of the gills on the legs of a crustacean.

The hindmost part of the body, the metasoma, represented by three somites in the embryo, is reduced to a small area in front of the anus. Behind this is the spearlike telson which helps to relate the king crab to the extinct Eurypterida.

King crabs lay eggs from which hatch interesting little creatures known as trilobite larvae from a superficial resemblance to the extinct Trilobita. At first they have no spine. They swim more actively than do their parents, and at intervals bury themselves in the same way. They grow by moulting; the spine appears at the first casting of the skin and gradually increases in length.

The affinity of Limulus to the Arachnida was pointed out by Ray Lankester in 1881. This inspired much work and discussion among phylogenists who found it difficult to decide whether or not Limulus and its allies, the Merostomata, were descendants of the Trilobites, and whether or not scorpions were evolved from marine ancestors. After many alternative opinions, present belief tends to support an evolutionary relationship between the marine Merostomata and the terrestrial Arachnida.

CHAPTER TWENTY

ARACHNIDA

Arthropoda in which the body is divided into a prosoma and opisthosoma, which may or may not be connected by a narrow waist or pedicel. The prosoma carries six pairs of appendages, the opisthosoma is usually without appendages. The respiratory system, consisting of lung-books or tracheae or both, opens on the opisthosoma, and the genital orifice is always on the lower surface of the second opisthosomatic somite.

There are eleven orders, six of which are:

SCORPIONES

Arachnida with two median and three pairs of lateral eyes. The opisthosoma is clearly divided into a mesosoma of seven somites and a flexible 'tail' of five. The pedipalpi are large chelate weapons and the telson is a venomous sting.

OPILIONES

Arachnida with a pair of eyes mounted on an ocular tubercle and a pair of odoriferous glands in the prosoma. The opisthosoma is short and of nine somites only. The legs are long, sometimes very long. The male has an extrusible penis.

PSEUDOSCORPIONES

Arachnida with twelve opisthosomatic somites. The chelicerae carry a spinneret and produce silk. The pedipalpi are large and contain venomous glands.

ARANEIDA

Arachnida in which the opisthosoma usually shows no trace of segmentation and contains silk-producing glands opening at a group of spinnerets near the tip. The male organ is situated on the tarsal segment of the pedipalpi.

SOLIFUGAE

Arachnida in which segmentation is obvious in both prosoma and opisthosoma. The chelate chelicerae are large, powerful and directed forwards, with a flagellum in the male. The pedipalpi end in suckers. The legs of the fourth pair carry racket-shaped organs.

ACARI

Arachnida, often highly specialized, in which a division between the third and fourth pairs of legs divides the body into proterosoma and hysterosoma. All appendages are liable to great modification; many genera are parasitic.

An arachnid is always recognizable because its head is never separated from its thorax which carries four pairs of legs. The fore part of the body, the prosoma, is protected above by a tough carapace and below by a plate-like sternum, which is missing or hidden in some orders. On the carapace are the eyes, simple ocelli with smooth surfaces, from four to twelve in number. The opisthosoma of some orders retains segmentation in the form of hard tergites and sternites, though in some orders all trace of this is lost in the adults.

In front of the first pair of legs are two other pairs of appendages. The foremost of these are the chelicerae, which may be relatively small, as in scorpions, large as in harvestmen, or very large as in Solifugae. They may consist of two segments only, as in spiders, which pierce their prey with them, or of three segments, as in scorpions, which use them to pick up things. Behind the chelicera are the pedipalpi, a pair of limbs with six segments, which in some orders are simple leg-like tactile organs and in others, such as the scorpions, are powerful weapons.

The legs which follow are of seven segments ending in two or three curved claws. With its legs an arachnid not only walks, it may also dig or swim or grasp its prey. On the legs are spines or setae of various types which act as tactile organs and help to detect scents or sounds, that is to say they are chemotactic or sonotactic devices. Usually there are no appendages on the opisthosoma, and the spinnerets of spiders are an exception.

In general, the Arachnida are nocturnal predators, tending to hide during the day. With very few exceptions they are carnivorous. Like many land invertebrates they risk desiccation, which they overcome by suitable habits, by a choice of moist environments, and by the existence of a layer of wax in the cuticle. They keep their bodies clean and their limbs efficient by constant preening.

All arachnids, except scorpions which produce living young, lay eggs from which there hatch not larvae but nymphs. They grow by periodic ecdyses or castings of the exoskeleton, as do many Crustacea.

They are well adapted to fasting and many species can live for a year or more without food. They have also evolved an unusually elaborate system of courtship behaviour in which the sexes indulge when they meet as a preliminary to mating. Their behaviour in most circumstances is almost wholly instinctive and social organization is very rare.

SCORPIONES

Apart from their traditionally formidable nature, scorpions (Fig. 50) are of zoological interest as representatives of the earliest type of land arachnid. There is sufficient resemblance

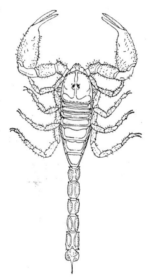

Fig. 50. Buthus, a scorpion

between them and the Eurypterida to suggest that ancestral scorpions lived in the sea. Fossil scorpions are very plentiful.

The two most conspicuous features of a scorpion are its large pedipalpi and its mobile tail with a terminal sting. The two parts of the body are jointed across their whole breadth and the prosoma shows no trace of segmentation, being covered with a uniform carapace. On this are the eyes, four to twelve in number. The opisthosoma consists of a mesosoma of seven somites followed by five narrower somites forming the so-called 'tail'. This can be raised in an arch and the sting thrust forward over the animal's head. Its venom is soon fatal to insects; its effects on human beings are varied, depending on the species of scorpion.

The mystery of the scorpion lies in a pair of comb-like organs, the pectines, attached to the lower surface, just in front of the lungs. They appear to be sense organs of some kind, but their real purpose has never been satisfactorily described.

Scorpions are unique among Arachnida for their young are born alive and not hatched from eggs. They spend the first few days of their lives on their mother's back. The number of eggs laid in a year is seldom more than fifty, a low reproductive capacity reflecting the relative absence of natural enemies in their surroundings.

Scorpions are found only in the warm or hot regions of the world, even including the comparatively barren desert areas. They burrow for shelter into the sand and are naturally protected against excessive loss of moisture by a layer of wax in the cuticle.

OPILIONES

Harvestmen (Fig. 51), being widely distributed and of reasonable size, are more familiar than scorpions. The body is usually oval, with only a groove marking the division between

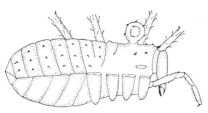

Fig. 51. Opilio, a harvestman

the two parts, and shallow grooves or rows of spicules marking the somites.

In most species there are two large eyes placed back to back

on a median turret on the carapace, and between this and the
border are the openings of a pair of odoriferous glands. The
secretion of a nauseous fluid from these is one of the harvestman's methods of self-protection.

The chelicerae, of three segments, end in forceps and the
pedipalpi behind them are leg-like. The legs are often long,
sometimes extraordinarily so; they carry many sensory setae,
the chief sense organs of the animal. They are of seven segments
and the terminal tarsus is often sub-divided into a number of
pieces. The legs end in smooth claws.

Opiliones are mainly nocturnal, predatory creatures, feeding
on a wide variety of insects and other small animals. They
constantly drink and occasionally suck the juice from fallen,
bruised fruit. The male possesses an extrusible penis, an
unusual feature in Arachnida, and the sexes mate freely on
meeting without courtship. The ovipositor is a long tube and
the eggs are laid in moist surroundings under ground.

PSEUDOSCORPIONES

Outwardly the members of this order resemble true scorpions without tails, for they have the same large chelate
chelicerae, the same smooth carapace and the same segmented
opisthosoma. But they are very much smaller, the largest
species being less than 6 mm in length.

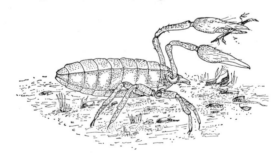

FIG. 52. A false scorpion

False scorpions (Fig. 52) are universally distributed, though
they are seldom seen, for they live under fallen leaves, loose bark
and in similar situations favoured by cryptozoic animals. When
brought into the open they move with a sedate, deliberate

138

tread until the outstretched palpi touch some obstacle, where-upon they dart backwards with the greatest speed.

A distinctive feature of false scorpions is their use of silk, which emerges from the small chelicerae in front of the mouth. They use this for building rest-cocoons, which they construct from bits of grit held together by silk threads and in which they moult, hibernate and lay their eggs.

The young ones are attached after birth to their mother by a short beak and are fed by an ovarian product, 'uterine milk'. They are purely carnivorous creatures and a few species have adopted the habit, known as phoresy, of grasping the legs of flies, beetles or harvestmen and thus securing easy transport. The stimulus for this action is unknown but hunger is suspected.

ARANEIDA

Spiders are by far the most numerous, the most varied and the most conspicuous of all the Arachnida, and they owe their biological successes to their ability to secrete and use silk.

The body of a spider is clearly divided by a narrow pedicel, the prosoma being covered by a smooth carapace on which lie the eyes, usually eight in number. In most spiders the dorsal shield is smooth and shows no traces of segmentation, but in some a groove appears to distinguish a head region from the rest, and in many the presence of muscles leaves a set of radiating marks. The opisthosoma is also quite smooth in most species, though there are a few that have formidable pointed spines variously arranged; but in the primitive sub-order Liphistiomorphae from the far east a full series of dorsal plates or tergites is retained.

The chelicerae are seldom conspicuously large and do not end in pincers. In most spiders they strike transversely and tend to meet in the middle of the transfixed victim, but in the large trap-door spiders they strike vertically and parallel to one another. Behind the chelicerae the pedipalpi resemble the legs that follow, except that in male spiders they show one of the most surprising modifications to be found among all the Arthropoda. The terminal segment is converted into an elaborate organ for transferring spermatozoa to the female. To do this it has first to pick up the spermatozoa from the abdomen, where the vas deferens opens on the lower surface of the second somite. This is beyond the reach of the pedipalpi, so that the

139

male spider spins a small sperm web, secretes a drop of semen upon it, and then charges its palpal organ with the fluid.

The fertilization of the female is usually preceded by a more or less elaborate series of courtship acts, but the myth that the female then kills and eats the male is not to be accepted as invariably true.

The opisthosoma has a pair of lung books near the pedicel, save in the trap-door spiders which have two pairs, and in a group near the tip there are normally four pairs of spinnerets. They secrete a protein liquid which hardens as it is drawn out, slightly different in composition from the silk of Bombyx mori, the silk moth, and used more widely. Most familiarly it is the material of which the web is spun, and the webs of spiders are of many different shapes and sizes, the spiral 'geometric' or 'orb-web' being the one that attracts most attention. The spiders that do not spin webs but hunt or lie in wait for their prey trail a drag line wherever they go, and silk is also used for tying up their victims, for making egg cocoons and hibernating chambers.

The venom glands of spiders are contained in the prosoma, their ducts leading to the tips of the chelicerae. The poison is quickly fatal to insects; some spiders are more virulent than others, and although the original Italian tarantula is not one of these, there are a few species, notably the Black Widows of the genus Latrodectus, whose bites may be dangerous and even fatal to man.

SOLIFUGAE

At first glance a Solifuge gives the impression of being a primitive type of arachnid, for segmentation persists in both halves of its body. There is a bulky swollen head in front, with two tergites behind it; on the abdomen nine tergites with corresponding sternites are visible. An obviously specialized feature is the large size of the chelicerae, directed forward from the head. They are sometimes almost as long as the rest of the animal, and the great strength of their crushing powers enables them to attack prey of many kinds. The pedipalpi behind them are leg-like and terminate in suckers instead of the more usual claws. They are useful for picking up food, for drinking and for climbing. The first pair of legs are longer and more slender than the others, and when the animal moves they are stretched out in front, acting as tactile sense organs.

The legs of the fourth pair carry mysterious objects known as malleoli or racket organs, whose function is unknown. The whole body is covered with long, sensitive setae giving it a distinctly fluffy appearance.

Solifugae are confined to hot countries and are the typical arachnids of deserts. They are pugnacious in habit and fight determinedly with scorpions or centipedes, but they are not venomous and do not attack man. There are some eight hundred species and their habits are imperfectly known.

ACARI

This order, which includes all the mites and ticks, is second only to the spiders in number of species and diversity of form. Their distribution is world-wide, including the arctic and antarctic regions into which even spiders have not penetrated as far as have the earthmites. Their life histories are varied and often remarkable and their economic importance is inferior among invertebrates only to that of the insects and Protozoa.

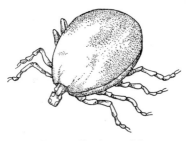

FIG. 53. Ixodes, a tick

Withal they are diminutive creatures, usually under a millimetre in length. The division of the body is inconspicuous, their mouth parts are varied and their legs end in claws, suckers or setae. Many species hatch as six-legged nymphs.

The largest and most familiar of the Acari are the ticks of the sub-order Parasitiformes. The sheep tick, Ixodes ricinus (Fig. 53), is often a serious pest of domestic animals. Its eggs are laid on the ground and the nymphs climb the grass and wait until an animal approaches, when, if possible they seize one of its hairs. Gorged with blood sucked from its host the tick drops to the ground and digests its meal, after which it moults and repeats the process until it is mature.

In the sub-order Trombidiformes is included the harvest-bug as well as the red mite, Tetranychus telarius, which is such a pest to fruit growers, sucking sap from the leaves and covering them with silk threads. The mite that is the cause of Isle of Wight disease in hive bees is Acarapis woodi and it also belongs to this group. It is so minute that it can live and breed in the trachea of the bee, which is suffocated as a result. A human parasite is Demodex folliculorum, which lives in the hair follicles and sebaceous glands of the skin where its presence in 'blackheads' has sometimes been described as universal.

The sub-order Sarcoptiformes includes the familiar cheese mites and sugar mites, but a more unpleasant member is Sarcoptes scabiei, the itch mite. This pest burrows in the human skin, laying eggs as she goes until a hundred or more are contained in a tunnel some centimetres long. The mite cannot escape, but the eggs hatch and are able to infect other people by direct contact.

In the sub-order Tetrapodili are the vegetable parasites that cause galls on leaves, big bud in currants and witches' brooms on trees; and the most surprising fact about this catalogue of acarine enemies is that, unlike insects, mites never seem to have been able to produce species that are of any benefit to the human race.

CHAPTER TWENTY-ONE

PYCNOGONIDA

Marine Arthropoda, in which the body is divided
into head or cephalon, trunk and abdomen. A
tubular proboscis projects from the head, which
typically also carries three pairs of appendages,
chelophores, pedipalpi and ovigers. The trunk
consists of four segments, each bearing a pair of
jointed legs. The abdomen is reduced to a
smooth oval sac.

Pycnogonida (Fig. 54) live on the sea bottom, from the
littoral zone to the greatest depths. Physical conditions here
are as nearly constant as possible all over the world, and perhaps for this reason Pycnogonida are of world-wide distribution.

A tubular proboscis, very characteristic of the group, projects forwards from the cephalon and ends in a triangular

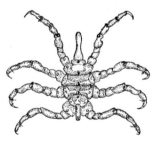

Fig. 54. Pycnogonum littorale

mouth with two lower and one upper lip. It is plunged into the
animal's food, which is sucked in through it.

Of the appendages, the chelophores are limbs of three or
four segments, the last two forming pincers. Behind them are
leg-like appendages, the pedipalpi, of eight to ten segments.
They are followed by the ovigers. There is nothing like these
in any arachnid, and the remarkable feature is that they are

present in most male Pycnogonida, while lacking in many females.

The possession of all three pairs of prosomatic appendages is far from being a constant feature. There are families without ovigers, others without chelophores, and others with neither chelophores nor pedipalpi.

The mesosoma or trunk consists of four cylindrical somites, with lateral processes carrying the legs. The shape of the body varies considerably, from elongated to almost circular. The existence of four pairs of legs would seem to relate these animals to the Arachnida, but that a number of species have five pairs and at least two have six pairs.

The abdomen is so reduced that it contains little more than the last part of the alimentary canal, and the gonads and the rest of the digestive organs are displaced into the near segments of the legs.

Eggs are laid through orifices in the second segment of the legs and hatch into nymphs, remarkable little creatures with chelicerae and three or four pairs of legs. Both eggs and nymphs are given a surprising amount of care by the male parents who carry both eggs and young about as they creep slowly over the sand. At intervals the growing young detach themselves and start independent lives.

They grow by ecdysis of the skin and as they grow the legs appear and simultaneously the nymphal pedipalpi and ovigers vanish, to reappear later in development. There are few similar instances of what seems to be an afterthought elsewhere in the animal kingdom.

Pycnogonida feed on the soft tissues of sea anemones, molluscs and Coelenterata, on which they are semi-parasitic. Sometimes they force their proboscis into the living victim, sometimes the chelophores tear off small pieces. Young pycnogonids may be found leading sheltered lives inside the hollow bodies of their hosts, but mostly they crawl very slowly over the sea bed, with occasional attempts to swim. Very little, however, is known of their habits.

Much of their zoological interest depends on the remarkable combination of arthropodan characters which they show, so that their position in any classificatory scheme is not easily determined. Their chelophores and their eight legs seem to relate them to the Arachnida; they resemble Crustacea in being marine, and the Caprellidae in particular in showing parental care in males. Like the Tardigradi they have no respiratory organs. They have been well described as having made an anthology of arthropodan characters, and if this be one of the ways in which evolution has come about, the Pycnogonida deserve close study.

CHAPTER TWENTY-TWO

MOLLUSCA

Triploblastica whose soft tissues are enclosed in a calcareous shell. The body is divisible into head, foot and visceral mass; the skin of this mass is extensive and forms a mantle which secretes the shell. Evidence of segmentation is confined to one class, and vestiges of the coelom form the pericardium. The larva, when present, may resemble the trochosphere of Annelida.

There are six classes:

MONOPLACOPHORA

Primitive Mollusca with a conical shell to which the body is attached by six pairs of muscles, segmentally arranged.

AMPHINEURA (POLYPLACOPHORA)

Mollusca in which the foot is flattened and the shell made up of a series of contiguous plates. The symmetry is bilateral; mouth and anus are at opposite ends of the body.

GASTROPODA

Mollusca in which the head bears tentacles and the body is often twisted so that the anus occupies an anterior position. The shell is consequently spirally coiled.

SCAPHOPODA

Mollusca in which the foot is vestigial and the shell is a tube open at both ends.

LAMELLIBRANCHIATA

Mollusca in which the body is compressed laterally and is enclosed in a bivalve shell. Symmetry is bilateral. The head is much reduced and the foot is hatchet-shaped.

CEPHALOPODA

Mollusca in which the foot surrounds the mouth and is drawn out into a number of prehensile tentacles. The mantle-cavity contains a syphon which expels water. The shell is reduced and is internal. There are large and complex eyes.

The Mollusca form a very large phylum, in which there have occurred adaptations to varied conditions, reflected in considerable modifications of the highly characteristic body. Thus the shapes and sizes of all three parts, head, foot and visceral mass, vary widely in the different classes.

The most constant and most characteristic feature of all Mollusca is the existence of a single or double fold of integument covering the visceral hump and known as the mantle. The cells of this mantle secrete the shell, which protects the soft bodies of all mollusca except the squids and octopuses, and by its weight puts them among the slowest-moving animals in the world. Between the mantle and the viscera is a space, known as the mantle-cavity, in which lie the gills (ctenidia) of the aquatic species, or which may function as a lung, as in the terrestrial snails.

In the shell three layers are distinguishable. The outermost, known sometimes as the periostracum, consists of a hornlike substance, conchiolin, variously coloured. Beneath it lies the prismatic layer, composed of crystals of calcium carbonate, similar in shape either to calcite or arragonite. Below and nearest the mantle is the nacreous layer.

Like nacre, pearls are the product of the mantle and the finest are of pure nacre throughout, made by oysters in warm or temperate seas. If a foreign body or a small organism, usually one of the parasites from the gut of a cartilaginous fish, finds its way between the shell and the mantle, the mantle is stimulated to cover the invader with nacre secreted by its cells. The commercial value of a pearl depends on its colour, size and shape. Many other bivalves can and do produce similar reactions to irritation, including the common freshwater mussel, but their creations have not the beauty or the glamour to give them great value.

The molluscan head carries one or two pairs of tentacles which are primarily sensory in function. There is in general no very definite line of demarcation between head and foot. The foot is essentially a muscular organ adapted for the slow creeping movement popularly associated with the snail, and for exerting an adhesive force, so obvious in limpets. Its action, which can be witnessed when a snail or a slug creeps over a sheet of glass, consists of a series of waves due to coordinated contractions of the longitudinal muscle fibres within. On land, movement is assisted by the secretion from a slime gland which opens just below the mouth.

The visceral hump was at first a conical mass containing the internal organs system. It has changed considerably in certain groups, especially in the Gastropoda where its spiral coiling has necessitated much rearrangement of the organs. Some of these modifications will be mentioned later.

The first part of the alimentary canal, immediately within the mouth, is the buccal mass. The mouth has a horny semi-circular upper lip, and below this, most characteristically, the radula. A radula is a strip of horny material, secreted con-tinuously from a radula sac below the buccal cavity. It is covered with transverse rows of minute backward-directed teeth, converting it into an efficient rasp with which the mollusc scrapes up its food. The form and number of the teeth vary so greatly that the comparison of radulae under the microscope has fascinated many malacologists, who have des-cribed some radulae with only three teeth in a row, while in others, such as the species Umbraculum, there may be three-quarters of a million teeth. Behind the buccal mass come the oesophagus and the stomach into which opens the hepato-pancreas, secreting amylases or other enzymes according to the food of the animal.

The heart consists of a median ventricle and normally two auricles contained in a pericardium, a vestige of the reduced coelom. The vessels associated with the heart often open into a large haemocoele. The blood is normally colourless.

Mollusca, being chiefly aquatic creatures, oxygenate their blood by gills or ctenidia. A ctenidium is rather like a double comb, a central axis with a row of delicate branches on each side. In the different classes the arrangement is considerably modified.

The nervous system is well developed. There is a circum-oesophageal collar connecting cerebral and pedal ganglia, from which nerve cords run to the essential organs. The loop-like pattern becomes twisted in some orders (Streptoneura) and again untwisted in others (Euthyneura).

The Mollusca include both bisexual and hermaphrodite species, the common snail being one of the latter.

During development the mesoderm originates in the same way as the Annelida, and the coelom arises by a separation of some of its cells, but the embryonic mesoderm shows no trace

148

of segments. This is the chief reason for the belief that Mollusca have descended from unsegmented ancestors, different from those of the Annelida. The existence of the molluscan trochosphere (Fig. 55) tends to cast a doubt on this hypothesis.

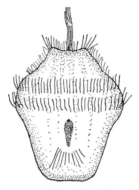

FIG. 55. Trochosphere of a mollusc

The eggs of most aquatic species of mollusc hatch as trochosphere larvae which have developed from the zygote by a process almost exactly similar to that of the polychaete annelid. In some families the trochosphere changes to a veliger larva, the internal structure of which closely resembles that of an annelid with additions.

MONOPLACOPHORA

This class contains the most primitive and in some ways the most significant of the Mollusca, well exemplified by Neopilina. Ten living specimens of this genus were dredged in 1952 from a depth of nearly 4000 m at a spot 200 miles west of Lima. The genus was supposed to have been extinct for some 500 million years. It was found that Neopilina had a head of three somites and a body of five, as well as five pairs of gills, five pairs of excretory organs and five cardiac ostia.

AMPHINEURA

Almost as primitive as the above, the Amphineura are well represented by Chiton cinereus (Fig. 56), the sea-mouse, in

149

which the mantle covers the whole upper surface and produces eight plates with an unmistakable appearance. The body is flattened and there is a row of ctenidia along each side. The

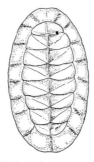

FIG. 56. Chiton cinereus, the sea-mouse

mantle covers the head, which is without eyes or tentacles. There is a radula and a simple nervous system, with the ganglia distributed along the nerve cords.

GASTROPODA

In every order of this class the structure is largely determined by the condition of the visceral hump. It appears that this hump, originally an upright cone, has become gradually changed into a spirally coiled mass usually lying on the animal's right side and enclosed in a single shell of a form familiar to everyone. This bodily change, described as torsion, has occurred to different extents in the different orders, the internal organs of which have, therefore, not all been affected to the same degree.

The class includes periwinkles, limpets, whelks and many others whose shells are familiar on our beaches; Limnaea and Planorbis in fresh water; and the snails which have exchanged their gills for a lung, as have the slugs whose shell is a vestigial internal plate. Here and in the rest are still to be found such features as the radula, the muscular creeping foot, the head with its tentacles, two auricles and two kidneys, all of which belong to the primitive molluscan organization.

The complication of torsion seems to have arisen from the gastropodan method of digestion, which tended to be concentrated in a dorsal diverticulum of the gut ending in the hepato-pancreas. This diverticulum grew longer until it fell sideways, and because one side grew faster than the other it became a spiral, with the result that the ctenidia are directed forwards, the auricles lie behind the ventricle, and the mantle cavity and anus open just behind the head. The advantages of torsion are obscure; it seems, however, to result in better filling of the mantle cavity, since both the movement of the animal and the action of the gill cilia would assist in the entry of water.

The Gastropoda most familiar to the laboratory zoologist are the only ones that have left the water and come to live on the

FIG. 57. Helix, a snail, the shell removed

land. These are the snails (Fig. 57) and slugs, air-breathing and hermaphrodite genera, which lay large eggs well supplied with albumen, and do not pass through a larval stage. The mantle cavity has become a lung, the periodic opening and shutting of whose circular orifice can be seen in every living snail.

The reproductive system is complex. The hermaphrodite gland is in the tip of the spiral from which a duct runs forward, receiving nourishment from the albumen gland for the gametes. The duct divides and a female portion, the vagina, reaches the mantle edge, having been joined by the ducts of the mushroom-shaped glands, the spermatheca, and the dart sac. The male half carries a long flagellum which becomes a penis. In copulation, each partner thrusts its dart into the other, where it usually remains, and each passes spermatozoa to the other. The sperm are likely to remain in the spermatheca until the ova leave the ovary, by which time the male portion of the ovo-testis is no longer active, an arrangement that makes self-fertilization impossible. The eggs are laid underground in late summer and hatch in about four weeks.

Most Gastropoda hibernate in a state of suspended animation. The opening of the shell is closed by an epiphragm and the rate of heart-beat drops to about half its normal value.

Slugs have lost both hump and shell, and so are sensitive to the water-content of the air, and continued dryness is fatal. They are accordingly strictly nocturnal, and perhaps their most surprising features are their abundance and the wide choice of their diet.

SCAPHOPODA

Although this is a small class it deserves mention because it stands in some ways between the Gastropoda and the Lamellibranchiata. The shell is a nearly straight and almost uniform tube, from the lower end of which the foot projects. The foot is pointed and is less used for crawling than for burrowing in the mud or sand, leaving the open upper end of the shell just exposed. Water enters and leaves the mouth cavity through this opening. There is a radula, but no sense organ on the head; there are no ctenidia and no pulsating heart. The most familiar species is probably Dentalium vulgare, the elephant tusk shell.

LAMELLIBRANCHIATA

This class contains the molluscs popularly known as bivalves, because their bodies, which are laterally compressed, are protected by a shell in two pieces joined by a hinge. The two halves are closed firmly by two powerful adductor muscles. Included in the class are the scallop, Pecten; the sea mussel, Mytilus; the fresh water mussel, Anodon; the oyster, Ostrea and the shipworm, Teredo.

Throughout the class the head is reduced to a vestige and there are no eyes, tentacles or radulae. A pair of labial palps mark the position of the mouth between them. The foot is not flat but is wedge- or hatchet-shaped. It is well supplied with blood vessels, the entry of blood into them producing a turgescence of the organ and enabling it to plough its way through the mud or sand. In this way the swan mussel is said to be able to move a mile a year.

The most characteristic feature of the class is, however, the enlargement of the gills and their taking a share in the capture of food. The form of the body between the two halves of the shell allows room for the axes of the ctenidia to extend along the whole side of the mantle cavity (Fig. 58). The 'teeth' of the

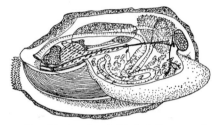

FIG. 58. Anodon, half of shell removed

ctenidial combs become much longer, hanging down to the bottom of the cavity, turning and ascending again, and, by joining laterally, producing gill plates. Each plate thus has a descending and an ascending lamella. At the front end of the foot the gill plates are attached to it; at the hind end the lamellae divide the mantle cavity horizontally, the two portions opening to the exterior as dorsal syphon and ventral syphon.

Cilia on the gills produce a constant stream of water which enters at the ventral and leaves by the dorsal syphon. Thus the gills have come to carry out the function of filter-feeding, and respiration is directly effected by the mantle surface. The food particles are entangled in mucus and driven towards the labial palps and the mouth.

The alimentary canal consists of oesophagus, stomach, intestine and rectum. The intestine makes one or two loops in the foot, passes through the pericardium and ends at a rectum. The system is supplied with an amylase-like enzyme, enabling the animal to digest carbohydrates, a peculiarity which exists also in some of the Gastropoda.

Most members of the class are bisexual, but Pecten and Ostrea are hermaphrodites with sex glands which function alternately as producers of ova or sperm. From the eggs the usual product is a trochosphere larva, followed by a veliger.

Some species retain the developing embryo within the ctenidia. A peculiarity is shown by members of the family Unionidae, the larva of which is known as a glochidium. These are liberated from the shell of their mother as a fish swims above her; they attach themselves by a thread or byssus to its body and remain there, encysted and parasitic, until they break away as small mussels.

CEPHALOPODA

This extremely interesting class contains the beautiful pearly nautilus, the cuttle fish of which Sepia is the most familiar, the octopus and its allies.

The striking external feature is the conversion of the head and foot into a number of active tentacles; some of these are of great muscular strength, are provided with suckers and their potential danger to divers is well known. More interesting than this, zoologically, is the development of their nervous system, and in particular the development of large eyes unbelievably similar to the eyes of mammals. The eggs are well supplied with yolk and hatch directly with no larval stage.

Of the three examples mentioned, Nautilus is the only one that lives in a shell, spiral in form and of considerable size, with the animal living in the last and largest chamber. The other parts of the spiral are those which it occupied earlier and abandoned as it grew and secreted a fresh refuge. All the chambers are connected by a thin tube, called a siphuncle, which is filled with blood and secretes air into them, making them so buoyant that the mollusc can swim easily. Sepia, the cuttlefish, has an internal shell, a white calcareous ovoid, often found on British shores. The octopus has no shell.

The tentacles of Nautilus are numerous and arranged in two rings round the mouth. They have no suckers but they are sticky, and with their help Nautilus drags itself over the rocks at the bottom of the warmer seas. The foot is thickened where it comes into contact with the shell, and when the animal withdraws itself this part acts like an operculum. Another modification of the foot is seen in a pair of lobes forming a funnel, and the contraction of the muscles forcing out a stream of water enables the animal to swim gently.

The tentacles of Sepia are ten in number, eight short and two long. Since there is no external shell the funnel is relatively large and the mantle mobile. The edge of the mantle becomes a fin and helps the squid to swim. Water is alternately drawn into and driven from the mantle cavity through the funnel, the normal process of respiration, and in moments of stress a rapid current is expelled and the creature is driven backwards by jet propulsion. At the same time there may be an ejection of black ink, which, clouding the surrounding water, helps the squid escape from danger. Nautilus and the octopuses have no ink sac.

The long tentacles of Sepia, called hectocotylus, play a remarkable part in reproduction. The spermatozoa are enclosed in elastic tubes, spermatophores, which pass out through the funnel on to one of the hectocotylized tentacles. This is adapted for transferring the spermatophore to the female, where it breaks open and liberates the gametes. In some of the octopods the modified tentacle is thrust into the mantle cavity, detached from its owner and left in position. It was at first described as a parasite under the name of Hectocotylus.

The giant squid, Architeuthis princeps, has tentacles which span forty feet and make it the largest invertebrate in existence.

The pearly nautilus has a relatively simple eye, consisting only of a pit with the retina at the bottom and no lens, a condition which is probably related to its nocturnal habits. The eye of the squid, on the other hand, has a cornea with eyelids in front and an iris and a lens behind it. The lens is supported by ciliary muscles, and a retina completes the resemblance to the eye of a mammal. Taken in company with other features of these Mollusca, the eyes suggest to us that the Cephalopoda may be regarded as the most highly evolved of the invertebrates.

A high degree of evolutionary advance argues a long evolutionary history, and this fact is indeed one of the features of Nautilus. It is the only cephalopod that lives in a shell, the descendant and the only survivor of a great number of species that flourished between the Cambrian and the cretaceous periods. The beautiful spiral fossils, often of considerable size, and familiarly known as ammonites, are evidence of their dominance as the marine invertebrates of the Mesozoic era.

PHYLUM BRACHIOPODA

As a postscript to the Mollusca, the phylum Brachiopoda may be mentioned. Often known as 'lamp shells', they are unsegmented and coelomate, and are covered with a bivalve shell which may or may not be hinged. All are marine and occur both in shallow and deep water.

The valves of the shell are placed dorsally and ventrally, the hinge is posterior and an efficient muscular system opens and closes the two parts. The shell consists of an outer periostracum covering crystalline calcium carbonate.

Mouth and mantle cavity open towards the front. Below the mouth is a row of protrusible tentacles, provided with cilia and forming the main part of the characteristic feeding organ known as the lophophore. Cilia on the tentacles produce two inward currents of water at the sides and an outward current between them. The mouth leads to a stomach into which a digestive gland opens. In some genera the intestine ends blindly.

The coelom is large, extending into lophophore, tentacles and mantle. It contains a pair of nephridia with large nephrostomes through which the gametes are also liberated. The sexes are separate and there is a larva which vaguely resembles a trochosphere.

Although extremely plentiful in the littoral zone of some tropical shores, the Branchiopoda are of little importance. They have, however, one outstanding feature. The species Lingula pyramidata, which occurs in the seas off China, Japan and Australia, seems to resemble in every detectable detail its ancestors whose shells were preserved in the Cambrian rocks, the oldest of all fossiliferous strata. Lingula proclaims the fact that although evolution is universal it is not inevitable.

CHAPTER TWENTY-THREE

ECHINODERMATA

Marine triploblastica, whose bilaterally symmetrical larvae develop into radially symmetrical adults, usually with five 'rays' projecting from a central disc. The coelom consists of a perivisceral portion communicating with the exterior through a dorsal pore or madreporite, a water-vascular system and perihaemal and gonadial portions. Neither blood nor excretory system is present. A calcareous skeleton exists in the mesoderm. Locomotion is effected slowly by means of characteristic tube-feet. The nervous system is diffuse. The sexes are separate.

There are five classes:

ASTEROIDEA (starfishes)

Echinoderms with a star-shaped body and tube feet from ambulacral grooves in each ray. The larva is a bipinnaria.

OPHIUROIDEA (brittle stars)

Echinoderms with a clear distinction between disc and arms. Ambulacral groove closed. Larva is an ophiopluteus.

ECHINOIDEA (sea urchins)

Echinoderms with a globular body covered by a test of closely-fitting plates beset with spines.

HOLOTHUROIDEA (sea cucumbers)

Echinoderms with an elongated body with skeletal plates and no spines. Some tube-feet form tentacles round the mouth. Larva is an auricularia.

CRINOIDEA (sea lilies)

Sessile, stalked Echinoderms, the five arms bifurcated.

If to be unique is also to be interesting, then the Echinodermata may claim to be among the most interesting animals that a zoologist can meet. Fundamentally their uniqueness is due to one dominant characteristic, the radial symmetry of their body, which develops from a bilaterally symmetrical larva.

Echinoderms are wholly marine animals, and the common starfish of our shores is by far their most familiar species. It is evident at once that five portions, arms, radiate from a central portion; and a momentary examination of a starfish on the beach shows that its upper surface is moderately hard and the undersides of its five arms carry two rows of very peculiar organs, known as tube-feet. It is also obvious that the mouth occupies the middle of the lower surface.

The first modification that must be made in this simple observation is the correction of the apparently obvious terms upper and lower, for these surfaces are derived from the right and left sides of the bilaterally symmetrical larva, and have taken up their final positions at the time of metamorphosis. The terms oral and aboral may be used, in which connection it may be pointed out that in the stalked, sessile Crinoids the mouth is directed upwards, in the Holothurians horizontally, and in the rest downwards.

The tube-feet are characteristic and are confined to the phylum. They are flask-shaped bodies, generally arranged in rows and in communication with a water-vascular system. Changes in the pressure of the fluid within cause their ends to act as suckers when the animal walks. Starfish use their powers of suction to pull open the shells of bivalves, and probably in all classes the tube-feet assist in respiration.

The ossicles, which give the epidermis of a starfish its toughness, are flattened spines, present in more pointed form in other classes. In some orders they assume a specialized form and are known as pedicellariae. These are small pincers, apparently used in cleaning the body. Another characteristic of the Echinodermata is the madreporite. This is a perforated plate on the aboral surface and through it sea water enters the water-vascular system to every part of the interior, including the tube-feet.

For the rest it may be said that in general the alimentary canal is very variable in the different classes, sense organs are poorly developed, and there are no excretory nephridia.

ASTEROIDEA

The ossicles of the outer layer are not united and two rows of them are placed along the sides of each arm. Spines arise from many of these plates, pedicellariae among them.

The mouth leads to an oesophagus and a stomach, above which is a pentagonal storage sac, giving off five pairs of pyloric caeca to the arms (Fig. 59). These caeca secrete enzymes as

FIG. 59. Section through arm of a starfish

well as acting as storage vessels. The pyloric sac leads to a short rectum. There are ten groups of gonadial tissue, with two ducts at the base of each arm. In feeding, the whole stomach may be everted through the mouth and wrapped round the prey which is digested outside the body.

OPHIUROIDEA

The arms are attached to the lower surface of the disc, are more slender than those of the Asteroidea and are easily broken off. Plates below the arms cover the ambulacral groove, which thus becomes a tube and is known as the epineural canal. Large ossicles inside each arm, functioning almost like vertebrae, prevent the existence of pyloric caeca. The tube-feet have no suckers. The stomach cannot be everted.

ECHINOIDEA

The sea urchin, Echinus esculentus (Fig. 60), is derivable from the starfish by drawing the arms into the body and covering the whole almost hemispherical object with a shell made of twenty rows of firmly united dermal plates. All plates are knobbly and carry spines and pedicellariae.

The mouth lies in the middle of the flattened base. At the

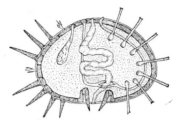

Fig. 60. Echinus esculentus, a sea urchin

opposite side of the hemisphere, the aboral side, the ends of the radii surround the central anus, while between every two radii there is a genital plate carrying the opening of the gonaduct. Unlike starfishes, urchins feed mainly on seaweed.

HOLOTHUROIDEA

The sea cucumbers are very different from the other Echinoderms, since they occur at all depths of the sea instead of living on the bottom. They are elongated and cylindrical in shape, and their pentagonal formation is far from obvious. The mouth is at one end of the cylinder and the short cloaca at the other, an almost normal arrangement. The alimentary canal consists of oesophagus, stomach, intestine and cloaca.

Their skeletal plates are very small, and instead of providing a protecting surface they are buried in the body wall. Moreover, there are no spines, so characteristic of the other classes, and no pedicellariae. The tube-feet, too, are reduced, save in the region round the mouth, where they are larger than might be expected, and form tentacles which pick food from the sand and push it into the mouth.

A peculiar method of respiration depends on two much-branched diverticula from the posterior part of the gut. Water is driven into these by contraction of the cloaca and brings oxygen to the coelomic fluid. In some genera these respiratory tubes give rise to a number of adhesive outgrowths, known as Cuvierian organs. When the animal is stimulated these are forced out through the cloaca, elongate in the water, and entangle the enemy.

Interesting as is the organization of so advanced an animal in a radially symmetrical form, the chief theoretical problem presented by the Echinodermata is that of their evolution. This cannot even be outlined without reference to the large number of fossil forms that have been described, and to do this would be beyond the scope of this book. In the briefest of terms the story seems to be somewhat as follows.

The earliest Echinodermata were the sessile Eocrinoidea, ancestors of the sea lilies of today. This class was in the majority (was 'more successful') in the Mesozoic, and gave rise to Edrioasteroidea, the ancestors of the starfishes. At this time the Echinodermata had depended for food on such organic matter as sank through the water, but a remarkable change occurred when the Edrioasteroidea turned over, used their tube-feet for walking, and began to walk about in search of food. It is to be noted that all present day classes seek food in different ways.

Following the above considerations, there arises the question of the relation of this peculiar phylum to the other phyla of the animal kingdom. Emphasis here falls on the larvae, and in particular on the Auricularia larva found in the Holothuroidea and to a lesser extent on the Bipinnaria larva of the Asteroidea. Both these show resemblances, which must be more than fortuitous, to the Tornaria larva of Balanoglossus, one of the Protochordata. If this is regarded as significant it suggests that the Echinodermata may be the invertebrate group from which the Chordata arose. On the other hand, Balanoglossus and the other Hemichordata are rather distant relatives of the other groups of Protochordata (see Chapter Twenty-four), and the evidence for chordate ancestry in the Echinodermata may be little more than a hint, difficult to interpret and of doubtful value.

POGONOPHORA

As a postscript to the Echinodermata a few lines may be added mentioning their apparent relatives, the Pogonophora.

These are sessile animals leading inconspicuous lives at the bottom of the ocean, with a thousand or more fathoms of water above them. They have very long and very thin bodies with an array of tentacles at one end, and they permanently inhabit tubes made of chitin. They are almost incredible, for they are animals without mouths, alimentary canal or anus, deficiencies that are usually only to be found among parasites, but these creatures are filter-feeders. Cilia waft food particles into the hollows of their tentacles where they are digested. The embryo begins to develop a gut, but this disappears during its growth.

The first of these remarkable animals was discovered in 1914, and twenty years passed before another was described. At first believed to be a sort of Polychaete annelid, they were later raised to the rank of a phylum. They are now known to be very plentiful at great depths, especially in Antarctica, in which respect they superficially resemble the Pycnogonida. Over forty species have been described and placed in two orders.

PART FOUR
VERTEBRATE ZOOLOGY

CHAPTER TWENTY-FOUR

CHORDATA

Coelomate triploblastica in which there is an internal axial skeleton composed first of notochord but which may later be replaced or supplemented by cartilage or bone. The external skeleton, when present, is of scales, feathers or hair. The pharynx is at first pierced by a number of visceral clefts or gill slits, which may later close. The central nervous system is a hollow dorsal tube, anteriorly expanded in the higher classes into a three-lobed brain. Blood, contained in a closed vascular system, is driven forwards in a ventral aorta and backwards in a dorsal aorta. There are dorsally situated kidneys, and also a segmented part of the body posterior to the anus, the tail.

The group is one in which, because of its comparative youth and the large number of fossil remains, the course of evolution can be traced with reasonable confidence. This is displayed in a rather elaborate scheme of classification indicating, among other things, the successive appearance of skulls, jaws and paired limbs:

PHYLUM CHORDATA

Sub-phylum Acrania (Protochordata): Chordata with neither skulls nor jaws

Class Hemichordata: Acrania in which the notochord is confined to the proboscis

Class Urochordata: Acrania in which the notochord is found only in the larva

Class Cephalochordata: Acrania in which the notochord is continued forward to the tip of the snout

Sub-phylum Craniata (Vertebrata): Chordata possessing skulls

Infra-phylum Agnatha: Craniata having no jaws

Class Cyclostomata: Agnatha with permanently circular mouths

Infra-phylum Gnathostomata: Craniata with jaws
  Super-class Anamnia: Gnathostomata in which the
        embryo is not protected by an amnion
     Class Chondrichthyes: Fishes with cartilaginous
        skeletons
     Class Osteichthyes: Fishes with bony skeletons
     Class Amphibia: Anamnia with four pentadactyl
        limbs
  Super-class Amniota: Gnathostomata in which the
        embryo is enclosed in an amnion
     Class Reptilia: Amniota covered with dermal scales
     Class Aves: Amniota with feathered wings
     Class Mammalia: Amniota with mammary glands

Because of their size, wide distribution, diversity of habits
and the economic value of a number of their species, the
Chordata are much more familiar than the far more numerous
invertebrates and they are more often hunted, photographed
and described than the members of any other phylum.

We owe our knowledge of their past history largely to their
possession of bone. Whereas the bodies of most invertebrates
are either soft or covered by an exoskeleton that is tough and
impermeable rather than hard and durable, the bodies of
vertebrates are supported by an endoskeleton of bone or
cartilage or both. Bones, and especially the modification of
bone known as teeth, have often survived the passage of many
centuries, and are still to be dug up in sufficient numbers to be
of value to all zoologists. To mention only one result of this, the
classification of fishes has been completely re-written in the
light of recent palaeontological knowledge.

The body of a chordate is both metamerically segmented
and coelomate. Undoubtedly the most obvious illustration of
its metamerism appears when a fish is being eaten and its
muscles can be readily separated with knife and fork. This
segmentation appears first in the embryo, when a series of
mesodermal somites extends from the brain to the other
end of the body. This sets the pattern for segmental repetition of vertebrae, of lateral nerves and often also of blood
vessels.

The coelom arises as a division in the mesoderm and may,
therefore, be temporarily segmented. It comes, however, to
form the large body cavity or splanchnocoel of the animal, the
space in which lie the viscera, suspended on connective tissue

mesenteries. A part of the coelom becomes separated from the rest and surrounds the heart, forming the pericardium.

ALIMENTARY SYSTEM

The alimentary canal is a tube lined with epithelium that is largely glandular, running from mouth to anus. Its functions are the digestion of food, and in the anterior region only, respiration.

The mouth is characterized by the possession of teeth and a tongue, and receives the secretions of the salivary glands. It is followed by the pharynx, which is a more obvious region in the fishes than in other classes where the pharyngeal clefts remain open. The oesophagus, into which it passes, is no more than a conducting passage along which the swallowed food is driven by rhythmical muscular contractions known as peristalsis. It widens to form the stomach.

The stomach is a more variable organ, for it is often the first part of the canal in which the food is attacked chemically. Hence different diets need different stomachs. Generally speaking, the carnivorous animals possess the simplest stomachs: that of the frog is a good example. In graminivorous animals a division into cardiac and pyloric portions becomes more evident, and in those that chew the cud it may be a complex system of chambers and valves. Again, it may be supplemented, as in birds, by a muscular gizzard which helps mechanically in the trituration of the food.

The stomach leads to the duodenum where the chemical changes are usually completed and the food finally reduced to a soluble form. Here it receives the bile from the liver and the pancreatic juice from the pancreas. This gland is stimulated by the arrival of the hormone secretion, produced by the cells of the duodenum and carried round the body by the blood.

The function of the small intestine or ileum is the absorption of the products of digestion into the blood and lymph streams. It follows that the longer the intestine the greater will be the chance for complete absorption of its contents, and a coiled intestine, the total length of which is several times the length of the body, is found in all classes above the Chondrichthyes.

The ileum gives place to the large intestine or colon. At their junction there is a longer or shorter diverticulum, the caecum, which ends blindly at the vermiform appendix. Here are to be found the multitudes of symbiotic bacteria, the value of which is their ability to decompose the cellulose in the diet.

167

In consequence it is longer in graminivorous animals than in those that eat meat; and it may be absent altogether. This condition occurs in man, where only the vestigial appendix remains.

The colon is the region from which water is absorbed and returned to the body. Much water has been used in conveying the digestive enzymes from the glands to the alimentary canal, and the system cannot afford to lose it all. The fluid that enters the colon is reduced to about a tenth of its bulk by the time it reaches the end.

Finally the rectum is a muscular conducting tube, the walls of which have but a slight absorptive power.

VASCULAR SYSTEM

The blood vascular systems of the Chordata show a steady evolution (Fig. 61). In the Chondrichthyes the heart consists of a ventricle and a single auricle. Blood enters the ventricle from the triangular sinus venosus, and leaves the auricle by the truncus arteriosus for the ventral aorta. From this spring five

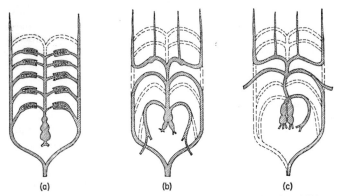

(a)   (b)   (c)

FIG. 61. Evolution of arterial arches. (a) Fish. (b) Amphibian. (c) Mammalian

pairs of afferent branchials which carry the blood to the gills for reoxygenation. It is collected from the gills in four pairs of efferent branchials which join the dorsal aorta. This distributes the blood forward to the head and backwards to the bodily organs, until, passing the cloaca it becomes the caudal artery.

The de-oxygenated blood collects in broad spaces known not

as veins but as sinuses. That in the head enters the nasal and orbital sinuses, and so, via the anterior cardinal sinuses, passes to the ductus Cuvieri.

From the tail the caudal vein divides into two renal portal veins which break into capillaries in the kidneys. In this way the blood, overloaded with toxic products of the muscular activity of the tail, is purified before it enters the rest of the vascular system. Between the kidneys arise the large posterior cardinal sinuses which join the ductus Cuvieri.

Blood from the liver travels in the hepatic portal vein and thence through the hepatic sinuses to the ductus Cuvieri. Lastly, blood from the fins enters iliac, brachial and subclavian sinuses which join the deep lateral sinus to the ductus Cuvieri.

This short account makes it clear that the blood travels in a single circuit, and that all blood in the heart is deoxygenated and is driven by the heart to the gills before travelling round the body.

Great changes occurred when the fishes gave rise to the Amphibia. The heart here consists of a single ventricle and two auricles. A triangular sinus venosus opens into the right auricle, and a short truncus arteriosus leads out of the dorsal side of the ventricle. From the truncus arteriosus arise three pairs of symmetrical arterial arches, representing a reduction of the afferent branchials of the fishes, which were themselves a reduction from the primitive number of seven. These arches, known as the carotid, systemic and pulmo-cutaneous arches, carry blood from the ventricle to the head, trunk and lungs respectively. There is a degree of asymmetry in the rising from the left systemic of the coeliaco-mesenteric artery, branches from which run to the intestine, stomach and liver. Thereafter the systemics unite, forming the dorsal aorta which divides into the iliac arteries in the hind legs.

The returning venous blood from the head enters the sinus venosus by two anterior venae cavae. Each of these is formed by the union of three veins, known as the external jugular, innominate and subclavian.

Blood returns from the hind legs in femoral and sciatic veins which unite to form renal portals to the kidneys. The femorals give off the pelvics, which join and form the anterior abdominal to the liver. From kidneys and liver the blood passes by renals and hepatics to the posterior vena cava and so to the sinus venosus. From lungs and skin the blood returns in the pulmo-cutaneous vein to the left auricle.

It follows from the above that when the auricles contract the ventricle receives both arterial and venous blood simultaneously. Formerly much ingenuity was expended in a description of partial separation of these; it is now known that there is but little mechanism for such separation and that most of the blood in an amphibian's body is mixed to a degree that would be fatal to a mammal.

The reptilian blood system is fundamentally distinguished from that of the amphibian by the fact that oxygenation of the blood through the skin has disappeared, and the lungs are the sole respiratory organs. At the same time there are changes in the heart itself.

There are two auricles and a ventricle which is partly divided in most orders and wholly divided in crocodiles. Two aortic arches arise from a short truncus arteriosus, give off two carotids and unite to form the dorsal aorta. Right and left subclavians arise from the right aortic arch, and from the dorsal aorta arteries run, as in the fishes and amphibians, to all the organs of the body. Blood from the head and trunk is returned via the sinus venosus to the right auricle in two precaval and one postcaval vein: these are the analogues of the anterior and posterior venae cavae. There are femoral and sciatic veins which, as in Amphibia, form an anterior abdominal vein to the liver, while blood in the pelvic veins passes by renal portals to the kidneys.

In birds and mammals the pattern established in the reptiles is conspicuously modified. The heart has four chambers, but there is neither truncus arteriosus nor sinus venosus. The arterial system is now asymmetrical, with only the right aortic arch retained in birds and only the left in mammals. From the aortic arch there arise the innominates, which give off the carotids and subclavians. The dorsal aorta continues to the tail, supplying the organs en route, and the returning venous blood travels in veins which normally lie close to the corresponding arteries. In birds the femoral veins give off small vessels to the kidneys, but in mammals there is no trace of a renal portal system.

NERVOUS SYSTEM

The central nervous system of a chordate originates in the outermost cellular layer or epiblast of the developing embryo, when a dorsal plate of cells rolls up to form the neural tube. The cells of which this tube is made ultimately take on the character of neurones or nerve cells, and axons or fibres. It occurs in Amphioxus in almost this state. In the higher classes there is a gradual enlargement of the anterior end to form the three primary vesicles which later become the fore-, mid- and hind-brain. These are variously elaborated according to the needs and mode of life of the animal. The fore-brain develops the olfactory lobes in front, the pineal stalk above and the important ductless gland, the pituitary body, below. To the mid-brain belong the optic lobes and to the hind-brain the cerebellum and medulla oblongata.

From the brain there arise the paired cranial nerves, ten pairs in fishes and amphibians, twelve pairs in reptiles, birds and mammals. The familiar summary of these is as follows:

1	Olfactory:	to nasal organ
2	Optic:	to the eye
3	Ocular motor:	to four muscles of the orbit
4	Trochlear:	to the superior oblique muscle of the orbit
5	Trigeminal:	to the nose and upper and lower jaws
6	Abducens:	to the external rectine muscle of the orbit
7	Facial:	to the head and neck region
8	Auditory:	to the ear
9	Glossopharyngeal:	to the pharynx
10	Vagus:	to heart, lungs and viscera
11	Accessory:	to larynx and neck
12	Hypoglossal:	to the tongue

This series of nerves at once suggests a segmental origin for the chordate head, which has lost all traces of its original somites in the course of evolution. Study of the development of the head of fishes has made it possible to reconstruct these changes and to detect eight anterior somites in the cephalic region.

URINO-GENITAL SYSTEM

The excretory and reproductive systems are in Chordata united in a way that is to be found in no other phylum of animals, and the description of the relations between the two sets of organs is one of the most characteristic features of chordate anatomy.

The excretory tubules of a vertebrate, analogues of the

nephridia of Amphioxus, arise segmentally, but not all at the same time. The pronephros comes first, followed by the mesonephros; and lastly the metanephros, where there is one, is formed as a separate organ. The following stages may be somewhat artificially distinguished.

First, there arise three tubules to form the pronephros. They open into the pronephric duct, which runs to the cloaca. This is a temporary condition, found only in the embryos of fishes and in larval Amphibia. In the higher chordates, the reptiles, birds and mammals, or amniota, the pronephros has only a fleeting existence in the embryo and disappears as the mesonephric tubules are formed behind it. The pronephric duct, however, does not vanish; it divides to form the Mullerian duct or oviduct, and the Wolffian or mesonephric duct.

The mesonephros is the functional kidney in the fishes and amphibians or anamniota. In the amniota it degenerates in the female, and in the male gives rise to the epididymis attached to the testis. The mesonephric duct carries both spermatozoa and urine to the cloaca. It therefore combines the functions of vas deferens and ureter and opens on a urino-genital papilla. The frog shows this condition well; the male dogfish shows an unexpected specialization, the elaboration of five separate urinary tubules leading into a duct of their own. In the female dogfish the mesonephric duct is the sole ureter and runs directly to a urinary papilla. In amniota the vas deferens, carried down to the scrotum, can be seen looping round the ureter on its way to the penis.

There is no metanephros in the anamniota. It is the kidney of all amniota, and its duct is the ureter, running to the bladder.

In addition to the systems of organs briefly described above there is in all Chordata a supporting skeleton of cartilage or bone or both. Every species of animal has the skeleton that best suits its needs, so that all skeletons are to some extent individual and different, but in a short account general outlines can be mentioned.

SKELETON

A skeleton is composed of an axial portion, the skull and vertebral column, and an appendicular portion, the limb-girdles and the limbs they support.

Skulls consist essentially of a cranium or brain box, with attached sense-capsules containing the olfactory, optic and auditory organs, and with the jaws variously attached to it.

172

The shape of a skull is determined by that of the brain, in which the increasing size of the cerebral hemispheres plays a major part. Thus skulls change from an elongated to an almost spherical shape in passing from fish to primates.

Jaw cartilages or jaw bones are variously attached. In fishes other than the Dipnoi the upper jaw is a rod of cartilage known as the palato-pterygoid-quadrate bar, which is joined to the cranium in front by the ethmopalatine ligament and at the back by a pre- or a post-spiracular ligament to the hyomandibula. This is a piece of cartilage that articulates directly with the skull. The lower jaw, or Meckel's cartilage, is also attached by ligaments to this hyomandibula. This arrangement is known as hyostylic jaw support, and is in contrast to the autostylic jaw support such as is found in mammals. Here the quadrate and the hyomandibula have become the incus and stapes of the ear, the upper jaw consists of maxilla and premaxilla fused to the cranium, and the lower jaw articulates with the squamosal bone.

In chordates with open, functioning gill clefts, a complex set or series of branchial arches strengthens the gill region. At first there were seven of these compound hoops, but later the first became the jaws and the second the hyoid apparatus close to the occipital region of the skull. The remaining five as found in the Chondrichthyes each consist of four cartilage bars, the pharyngo-, epi-, cerato- and hypobranchial on each side and a median basi-branchial below. In no fish are all five exactly alike nor is the complete set present. The epibranchials and ceratobranchials support the gills themselves.

The vertebral column consists of separate pieces which gradually replace the notochord. In Amphioxus the notochord alone provides the axial skeleton and support throughout life; in cartilaginous fish a vestige of it remains in each vertebra and fills the spaces left by the concavities at each end of the centra. Neural arches above the centra protect the nerve cord, lateral processes like short ribs project sideways, and in the tail region meet below to form the haemal canal, in which lie the caudal artery and vein.

In classes above this the proportion of notochord falls and the cartilage is replaced by bone. The vertebral columns of the Amphibia and Reptilia are extraordinarily diverse; the bird's column is characterized by the number of cervical vertebrae and the degree of freedom which they allow. The total number of vertebrae in a bird varies from 39 to 63. In mammals the

173

vertebrae fall into five groups, cervical, thoracic, lumbar, sacral and caudal, each with their own easily recognizable characteristics. The thoracic or dorsal vertebrae support the ribs, some of which join the sternum. In birds the sternum is the largest bone in the body, a great curved plate covering more than half the lower surface.

Among fishes limb-girdles are too rudimentary to deserve the description of girdles at all. In the other classes they follow the same basic design, and consist of a dorsal bone, the scapula

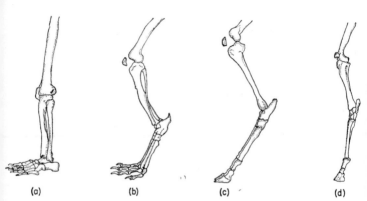

(a)   (b)   (c)   (d)

FIG. 62. Modifications of the pentadactyl limb
(a) Man. (b) Dog. (c) Sheep. (d) Horse

or ilium, and two ventral bones, the coracoid and clavical or the ischium and pubis.

A pentadactyl limb is articulated with the glenoid cavity of the scapula or the acetabulum of the pelvic girdle. The full set of bones is as follows, those of the fore-limb being named first in each case. A proximal bone, humerus or femur, is followed by two parallel distal bones, radius and ulna or tibia and fibula.

The wrist or ankle joint is based on a grouping of nine bones, arranged in a first row of three, a second row of one and a third row of five, an arrangement which is usually modified in any one group by fusion or omission of some of them. There follow the long metacarpals of the palm or the long metatarsals of the sole of the foot. Fingers and toes are supported by 2, 3, 3, 3, and 3 small bones known as phalanges.

Modifications of this plan (Fig. 62), as indicated above, can produce such a wide potentiality for adaptation to different environments and special functions that the pentadactyl limb is one of the most fascinating and versatile objects in all chordate anatomy.

CHAPTER TWENTY-FIVE

PROTOCHORDATA

Chordata in which the notochord is never replaced by cartilage, in which there is neither
brain nor skull, nor any heart, kidneys or paired
limbs.

There are three classes:

HEMICHORDATA (acorn-worms)

Marine, vermiform animals whose body is divisible into
proboscis, collar and trunk. The notochord is confined to the
proboscis, and since the anus is terminal there is no tail.

UROCHORDATA (sea-squirts)

Marine, sedentary animals whose larvae show chordate
characters, most of which are lost at metamorphosis. The
pharynx, much enlarged, is pierced by many slits and has become an accessory to ciliary feeding. The body is surrounded
by a test or tunic.

CEPHALOCHORDATA (lancelets)

Marine, free-swimming animals in which the notochord
extends to the anterior tip of the body. The pharynx has many
gill slits which open into an atrium. Excretion is by means of
nephridia.

It is obvious from the short diagnoses of these three classes
that they are neither clearly nor closely related, yet for convenience and by tradition they are often studied together,
largely in the hope that they may help us to understand the
origin of the vertebrate world.

HEMICHORDATA

The class of acorn-worms contains species whose true
affinities are, to say the least, obscure. Perhaps they represent
the most primitive form of chordate organization.

The familiar species Balanoglossus kowalenski is a worm-like animal about six inches long, its body divisible into proboscis, collar and trunk. The proboscis, which is variously shaped in different species, is supported by a notochord, the chief feature that gives Balanoglossus its right of entry into the Chordata, but it is not completely homologous with the notochords of Ciona or Amphioxus. These originate from the dorsal lip of the blastopore, while that of Balanoglossus is an outgrowth from the dorsal side of the gut, above the mouth.

Fig. 63. Larva of Balanoglossus

To this must be added the fact that Balanoglossus has no tail, one of the fundamental features of the chordates; its central nervous system lies on the ventral side of its body and not, as might be expected, on the dorsal side, while its blood flows forwards, not backwards, in its dorsal blood vessel.

Taken together, these features tend to set Balanoglossus and the Hemichordata apart from the other classes of the Protochordata, a fact which is of some significance for the following reason.

The tornaria larva of Balanoglossus (Fig. 63) closely resembles the larva of the Echinodermata, and on this has been based the hypothesis that the Echinodermata point the way to the ancestry of the chordate phylum. But if the Hemichordata are too unlike the other Protochordata to be unreservedly grouped with them, the value of the evidence for the hypothesis is greatly diminished.

The nature of these peculiar animals can best be understood by first summarizing their life history.

The fertilized egg develops into a larva which, something like a tadpole, seems to consist of body or trunk and tail. The mouth leads into a pharynx perforated in almost the normal way by pharyngeal clefts through which water escapes as the creature swims. There is an endostyle in the pharynx, which leads to oesophagus, stomach and intestine. The nervous

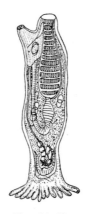

FIG. 64. Ciona

system is dorsal and hollow and swells in the front into a brain, in connection with which is an eye and a statolith. The tail consists of segmental muscles, supported by a notochord along its whole length.

At metamorphosis the larva settles down on its head, which is provided with suckers to fix it to the rock. It loses its tail and with it both notochord and nerve. The pharynx increases in size and the number of slits is multiplied many times: cilia produce a constant flow of water through the basket-like structure. The animal (Fig. 64) is now surrounded by a test or tunic of translucent tunicin, a substance allied to cellulose. The water from the pharynx escapes into the space between the body wall and the test. This space is the atrium, and the water finally leaves it by an orifice close to the inhalent opening. Faeces and gametes use the same route.

The adult possesses a few blood vessels and a heart which shows the unique property of periodically reversing the direction in which the blood flows. Of the nervous system only a large ganglion, representing the brain, remains and the animal is hermaphrodite, with both ovary and testis opening into the atrium.

Ciona intestinalis is the commonest British species and provides a good example of the so-called degeneration that may follow the assumption of a sedentary mode of life. Free-swimming Urochordata are also known and a number of colonial forms also exist.

CEPHALOCHORDATA

This is the best-known of the three classes, since it contains the much discussed Amphioxus or Branchiostoma, commonly known as the lancelet.

Amphioxus lanceolatus is found in shallow water off the European coasts wherever the bottom is sandy. It reaches a length of about two-and-a-half inches and spends most of its life buried vertically in the sand with only its fore-end protruding. Its popularity is due to the fact that it displays most of the basic features of the Chordata to those who are skilful enough to dissect it, and suggests the early stages of the evolution of the phylum to those wise enough to perceive them. Thus its notochord is a very obvious item in its body, serving as an axial skeleton and keeping head and tail apart when muscular contractions tend to draw them together. Above it lies the nerve cord, an important feature of which is that it has not expanded in front to form a brain, it merely forks and at the base of the bifurcation is a dark spot which is sensitive to light.

The whole body is covered with a cuticle of a single layer of cells, supporting a dorsal fin which runs without interruption from head to tail and passes round the tail as a caudal fin. There are no paired fins, one of the most obvious evidences of the primitive nature of the animal.

By contrast, Amphioxus has a very complicated mouth region. An oral hood fringed with oral cirri leads first into a buccal cavity, which contains a peculiar structure, the wheelorgan. At the back of this there is a partition, the velum, with a central hole which probably represents the mouth in the true sense of the word, since the pharynx lies directly behind it. A group of velar tentacles project backwards into the pharynx and act as a strainer.

The sides of the pharynx are pierced by gill slits, so called by analogy, for they have no gills. They are, however, many more than seven, which is evidently the primitive number in the true chordates, and this too is a specialized feature of the animal. Moreover, the gill slits do not open directly into the sea, but into a space known as the atrium formed by two folds of tissue which grow down from the dorsal region to meet below. The atrium opens through an atriopore, which instead of being median is on one side of the middle line.

The gonads are segmented and are unusually numerous, there being twenty-six cubical masses of gonadial tissue on each side. The gametes are shed into the atrium and escape into the sea, where fertilization occurs.

The most significant feature of Amphioxus is its excretory system, which consists of nephridia lying near the top of the gill slits and opening into the atrium. The inner end of each is a tuft of tubes ending in flame-bulbs like those of the Platyhelminthes, a most remarkable and unexpected sign of some sort of relationship between the Chordata and a phylum of invertebrates.

Clearly Amphioxus is a mixture of primitive and specialized features, as are, in fact, a number of other animals. The chief lesson to be learnt from it is the truth of the statement that there are many primitive organs but very few primitive organisms.

CHAPTER TWENTY-SIX

CYCLOSTOMATA

Primitive Craniata with a simple type of skull, with no jaws, and with a mouth that is permanently open. In the brain the pineal eye is recognizable and is sensitive to light: the pituitary body communicates with the exterior. The notochord is persistent, and the simple vertebrae have no centra. The olfactory organ is median and the ear has one or two semicircular canals. The kidney is a mesonephros. The larva, known as an ammocoetes, is a ciliary feeder with an endostyle.

There are two orders:

PETROMYZONES (lampreys), also known as Hyperoartii

MYXINI (hag fishes), also known as Hyperotreti.

The Cyclostomata, also called Marsipobranchii, bear a resemblance to fishes, and especially to the eels, which is only superficial, for the many differences point to the lampreys as being very primitive chordates, to be ranked just above Amphioxus.

A lamprey (Fig. 65), like Amphioxus, has a permanently

Fig. 65. A lamprey

opened mouth with no jaws, a persistent notochord, no scales and no paired limbs. There is a median fin round the posterior end of the body, but this fin has no supporting fin-rays, and there are seven conspicuous gill pouches on each side of the pharynx.

Internally, the skeleton is wholly cartilaginous. It consists of a very incomplete cranium partially protecting the brain, followed by a surprisingly large framework, the branchial basket, which supports the pharyngeal region. The notochord is surrounded by a sheath-like membrane. There are no vertebrae.

The mouth is very characteristic. A ring of cartilage holds open a large buccal cavity, which is well furnished with pointed, horny teeth. It contains the tongue, a muscular piston, which supplies the sucking action by which the lamprey feeds. Its method of feeding is to attach itself to the side of a fish and to rasp away its flesh, the teeth acting in much the same way as the radula of a mollusc.

Correlated with this process are the peculiarities of the respiratory system. Inside the short tubes which contain the gills are spherical gill chambers, which, by expanding and contracting, can draw water in over the gills and expel it again. In this way the gills are supplied with oxygenated water when the sucktorial mouth is sealed.

The brain is similarly constructed to that of the majority of Chordata, but cerebrum and cerebellum are very small. The spinal cord, covered only by fibrous tissue, is flattened rather than circular in section, and the dorsal and ventral roots of its nerves are not united. There is a single, median nasal sac, and in the ear the horizontal semi-circular canal is missing.

Cyclostomata are widely distributed, and are represented in Britain by three species, the sea lamprey (Petromyzon marinus) the lampern or river lamprey, Lampreta fluviatilis, and the brook lamprey, L. planeri. The first-named may reach three feet in length. They are unpopular with fishermen, whose bait they consume and whose catch they injure.

Sea lamprey swim up rivers in the spring for spawning. The eggs are laid on the mud or sand, and the exhausted parents drift downstream. The eggs hatch as larvae, known as Ammocoetes to zoologists and as prides or nine-eyes to others. They are blind, worm-like creatures that spend two years or more buried in the mud, leading a life reminiscent of Amphioxus. They have a continuous dorsal fin, but neither tongue nor teeth. At the top of the pharynx there is a velum, and inside there is an endostyle, also like Amphioxus.

The hag fishes, represented in British waters by only one marine species, Myxine glutinosa, are not greatly different from the lampreys in general structure. They differ, however, in their mode of life. Although much of their time is spent buried in the mud, they are voracious feeders and have been described as burying their pharyngeal regions in the flesh of the fishes on which they live as parasites.

Their most characteristic feature is a pair of rows of glandular pits along the whole length of the lower surface. In moments of danger a hag fish secretes from these pits masses of thread-like mucus in such quantity that they are described as turning the water into glue. This is apparently a protective reflex.

Hag fishes are periodic hermaphrodites, producing spermatozoa when they are young and ova later in life. Apart from these two peculiarities they may reasonably be regarded as degenerate lampreys.

CHAPTER TWENTY-SEVEN

CHONDRICHTHYES

Fishes in which the endoskeleton is wholly composed of cartilage, occasionally calcified, and in which the exoskeleton consists of placoid scales. The mouth and nares are ventral, and are externally connected by oro-nasal grooves. The upper jaw is the single palato-pterygoid-quadrate bar; the lower jaw is Meckel's cartilage; jaw-support is hyostylic. There is a spiracle and five gill clefts, not covered by an operculum, and a heterocercal tail. Paired fins are large. The heart has a well-developed conus arteriosus and the brain a large cerebellum; and there is a spiral valve in the intestine. There is no swim-bladder or lung. Fertilization is internal; the eggs are large, owing to the supply of yolk, and develop inside a horny capsule.

There are two sub-classes:

SELACHII (sharks, skates, rays and dogfish)

There is no operculum: a spiracle is present and the teeth are numerous.

BRADYODONTI (chimaeras)

The gills are covered by an operculum-like fold of skin: there is no operculum and the teeth are few.

The class Chondrichthyes contains some of the simplest of living fishes and also many with highly specialized features. Several obvious external characteristics distinguish them from the class of bony fishes.

The placoid scales are quite different. Each consists of bony base buried in the skin and carrying a backwardly directed dentine spike covered with enamel. These scales are the fore-runners of the teeth of all Chordata.

The mouth is not terminal but ventral and externally visible oronasal grooves connect it with the nares. There are several median fins on the back, supported by fin rays and in some genera preceded by a sharp spine. The tail fin usually has the dorsal lobe larger than the ventral lobe. The paired fins are large and act as stabilizers rather than as organs of propulsion. The beating of the asymmetrical tail has the effect of forcing the head downwards and so bringing the mouth close to the bottom-dwelling creatures on which the fish preys. The pectoral fins, in fact, control this downward thrust.

The pelvic fins of the male are joined to a pair of rod-like claspers with which the male graps the female, for among the cartilaginous fishes internal fertilization is the rule.

Lastly, and perhaps most obvious, is the presence on each side of the pharynx of five gill clefts out of which water that has entered the mouth is continually escaping. There is no bony plate or operculum to cover and protect these clefts. A migrant gill cleft with one vestigial gill, known as the spiracle, can be seen on the back of the head.

There is in all these and other fishes a more or less distinct lateral line running along the middle of each side. This lies over the lateral line canal which continues to the head. It contains numerous sense organs, known as neuromast organs, which appear to be the means by which the fish responds to the pressure of the water, the direction of the water current and the existence of vibratory disturbances in it. Their presence on the head may be shown by squeezing the head of a dead dogfish, when a drop of fluid appears at each organ. They occur only in fishes, cyclostomes and larval amphibians.

The sub-class Selachii is the larger of the two groups of cartilaginous fishes, and includes the two orders Pleurotremata, the sharks and dogfish, and the Hypotremata or rays.

Among the most interesting of the sharks are those in which the pharynx has six or even seven gill clefts and the back carries a single dorsal fin. They are among the most primitive forms of the class and are in sharp contrast to the active and efficient shallow water dogfish of our laboratories. This fish, Scyliorhinus caniculus, resembles other sharks in most respects, except size. The giant shark may reach a length of forty feet or more, and its strong jaws, armed with pointed saw-edged teeth, make it a formidable animal. It normally eats fish, but as is well known, may sometimes attack man. Most of the impressive sharks that find places in travellers' tales belong to the family of blue sharks, and the inexplicable hammer-headed shark is one of this group. Another species worthy of mention is the basking shark, which feeds wholly on small fry, filtered from its mouth as it slowly swims. When the supply of food falls below the amount needed to supply the energy to swim, the shark hibernates on the sea bed.

The skates and rays are broadly flattened fish, specialized for a life on the bottom. The gill clefts are under the head and water enters through the spiracle on the upper surface. The peculiar shape of the body is due to great enlargement of the pectoral fins. A remarkable species is the sting ray which has lost dorsal and caudal fins and reduced the tail to a lash armed with a few sharp spines.

The sub-class Bradyodonti, also known as the Holocephali, is a peculiar group resembling sharks more closely internally than externally. Thus they have four gill clefts which are covered by a kind of operculum. There are no placoid scales, and flat, studded plates on the jaws act as teeth. The dorsal fin is supported at its fore edge by a sharp spine, the ichthyodorulite, and in addition the males carry on their heads one or more spiked tubercles which are used in clasping the female.

Fishes must inevitably attract the attention of zoologists because they were the first successful group of chordate animals. Like nearly all the early stages of evolutionary progress, they were, and have virtually remained, aquatic organisms. It is of some significance that the uniformity of their environment is correlated with a close similarity in general outward form. The first consequence of this was the inclusion of all fishes in one class, Pisces; and if this were a monophyletic class then the general resemblance would be acceptable. The splitting of fishes into at least two living and one extinct class, is, however, an expression of the belief that the original Pisces was a polyphyletic group; and from this it follows that their outward resemblances must be a consequence of environmental influence. Such a deduction points decidedly in favour of the Lamarckian thesis that environmental circumstances determine the form of the organism.

The comparative abundance of fossil fishes has made it possible to construct a fairly trustworthy description of their evolutionary history. Some five hundred million years ago the sea was the home of the Ostracodermata, jawless, fishlike animals without fins. The first jawed fishes appear to have left the seas in favour of the rivers, where they throve and multiplied to such an extent that the rocks of four hundred million years ago compel us to call the Silurian epoch the 'Age of Fishes'.

From this point two streams diverged. In one of these the bony skeleton gave place to cartilage, or the cartilage failed to make the transition to bone, and the new group returned to the sea. Here they have remained and their descendants are still to be found, the sharks and dogfishes of the present day.

The alternative group of fishes retained their bone skeletons and met the increasing foulness and oxygen-deficiency of the water by developing lungs, along with two pairs of lateral fins. With this equipment two lines of possible evolution emerged, distinguishable as the ray-finned and the lobe-finned. In the former, the lung was converted into a swim-bladder, and its possessors became the ancestors of nearly all the bony fishes of today, including those that are now to be found in the fresh water of rivers and lakes.

The lobe-fins themselves split into two groups, one of which survives as the lung-fish or Dipnoi of the present. The other became the fringe-fins of Crossopterygii, and among these the Coelacanths spread all over the oceans by about seventy million years ago. The remaining Crossopterygii were more ambitious. Some of them returned to the rivers; others acquired nostrils through which air could enter the mouth on its way to the lungs, and finally, exploring the land, gave rise to the first Amphibia.

CHAPTER TWENTY-EIGHT

OSTEICHTHYES

Fishes in which the endoskeleton is wholly composed of bone and the exoskeleton consists of
flat bone scales. The mouth is terminal, the
external nares are dorsal. Jaw-support is variable.
The gills are usually covered by an operculum.
There is a swim-bladder, which may act as a
lung or as a hydrostatic organ. The tail is
homocercal, and there is seldom a spiral valve
in the intestine, which is elongated. The olfactory
lobes and the cerebellum are small. The eggs are
small and are laid in large numbers.

The classification of so large a group as this is likely to be
complex and uncertain: a practically helpful system, limited to
living species, is as follows:

SUB-CLASS CROSSOPTERYGII

Fishes in which the paired fins articulate with the limb
girdles by a single bone, giving mobility to the fin which is
covered at its base by a lobe of the body wall. Lungs are
present, adapting the fish to life in shallow or under-oxygenated
water. There are two orders:

Rhipidistia. Fishes related to the extinct Osteolepidoti, with a
body covered in characteristic scales. The tail is diphycercal and
the vertebral column projects between the lobes.

Dipnoi. Fishes with efficient lungs and a heart and vascular
system resembling those of the Amphibia. The scales and
skeleton are much reduced.

SUB-CLASS ACTINOPTERYGII

Fishes in which the paired fins have broad bases and are supported by horny fin-rays. The lung has become an air-bladder
or hydrostatic organ. There are four orders:
Chondrostei. Fishes in which the notochord is sheathed in
cartilage, neural and haemal canals exist, but no centra. There

is a spiral valve. There are five rows of bony scutes on an almost scaleless skin.

Polypterini. Fishes in which the dorsal fin is divided into a number of separate elements and the body is covered with shiny ganoid scales.

Holostei. Fishes in which a spiral valve, optic chiasma and conus arteriosus recall the Chondrichthyes. Biconcave centra are present as well as an air-bladder.

Teleostei. Fishes in which the skeleton is wholly of bone and the scales are thin plates of dentine.

With few exceptions the bony fishes have bodies of a fusiform shape which, shared as it is with so many other marine animals, is an example of adaptation supplemented by convergent evolution. The body of a bony fish is covered by dermal scales, thin plates of dentine embedded in sacs in the dermis and covered with epidermis. They grow with the growth of the fish, and exhibit annual rings which indicate its age.

There are dorsal and anal fins, usually supported by parallel fin rays. There are two pairs of paired fins, distinguished as pectoral and pelvic though they are not connected to corresponding girdles. In the course of evolution the pelvic fins have tended to move forward, and in some species are actually in front of the more stable pectorals. The tail is the chief organ of propulsion and is symmetrical in shape in most families.

The mouth is terminal. From the pyloric end of the stomach there is a group of caeca, among which lie the delicate tubes of the pancreas. The intestine is long and looped and has no spiral valve. Anus, ureter and genital duct have separate openings within a cloaca.

In the heart the conus arteriosus is merged with the ventricle and the ventral aorta is thickened, forming a non-contractile bulbus arteriosus. From the heart the blood travels to the gills. Each gill takes the form of two rows of triangular plates, one on each side of each septum between two gill clefts. Definite breathing movements occur. When the mouth is open the opercula lie close against the body and prevent the escape of water through the gill slits. When the pharyngeal passage is constricted the opercula lift and the water flows out while two pieces of bone near the maxilla prevent the loss of water through the mouth.

Above the stomach there lies in all familiar fish a silvery, hollow air-sac which opens into the oesophagus. This acts as a hydrostatic organ, enabling the fish by changing its density to rise or fall in the water. In other fishes it is a lung by means of which the fish can take in gulps of air and so oxygenate its blood.

In the brain there are small olfactory lobes, and the cerebellum is also moderate in size. Since the eyes are on opposite sides of the head there is no optic chiasma to unite their two impressions into a single picture. In the ear there is an otolith of bone, often of characteristic shape.

The kidneys have no connection with the testes. The tubules from each kidney open into an archinephric duct, and these unite to form a median ureter. Each testis has its own vas deferens, and inside the cloaca the rectum, the genital duct and the ureter open separately.

In contrast to the Chondrichthyes, fertilization is usually external and the eggs are unprotected. They are therefore very numerous, being numbered by the thousand at any one spawning. The 'fry' that hatch from them are often so different from the adults that they are technically larvae, for they may have a continuous fin round the body, a notochord and a functioning pronephric kidney.

RHIPIDISTIA

To this group belong the Coelacanthini, with the well-known Latimeria chalumnae (Fig. 66). This fish, believed to have been extinct for fifty million years, was represented by a single

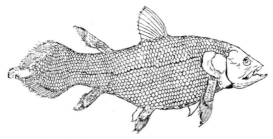

FIG. 66. Latimeria chalumnae, a coelacanth

specimen taken off Africa in 1939. Later, in 1952, a second specimen was caught and has since been followed by others which have received the attention they deserve.

They are large fish about five feet long, with characteristic overlapping scales. There is an unconstricted notochord, a swim-bladder that is not a lung, and a spiral valve in the intestine.

DIPNOI

The Dipnoi, or lung-fish, are represented by five living species, Ceratodus from Queensland, Lepidosiren from the Amazon, and three species of Protopterus from Africa. The young of all Dipnoi breathe by temporary external gills. The Australian species lives in foul water deficient in oxygen: at intervals it rises to the surface, gulps a mouthful of air, and at the same time emits a characteristic grunt, one of the earliest examples of a vertebrate voice. The body is covered with very large scales, whereas in the other Dipnoi the scales are smaller and the paired fins are whip-like rather than paddle-shaped.

The African and American lung-fish find the greatest value of their lungs during the heat of summer, when the water in which they live may be scarce or absent. They then aestivate. Protopterus buries itself in a capsule of hardened mucus, with a perforation to permit breathing; while Lepidosiren rests in a clay burrow, partly closed by a clay plug.

CHONDROSTEI

The sturgeons (Fig. 67) are marine fishes which spawn in rivers. Zoologically they are a remarkable mixture of the primitive, the specialized and the degenerate. They have a persistent notochord, a spiral valve and a spiracle, all primitive

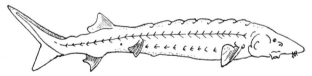

Fig. 67. Acipenser, the sturgeon

features; they have a peculiar specialization in the form of five rows of bony scutes or bucklers, each armed with a spine, and a bunch of four sensory barbels under their shovel-like snout. They have lost scales and teeth, save in the very young.

They are of value for their esteemed flesh, and they also provide caviare and isinglass.

POLYPTERINI

These are African fishes which, unlike the sturgeon, have retained their ganoid scales and possess a remarkable dorsal fin divided into 15 or more separate elements (Fig. 68). A primitive

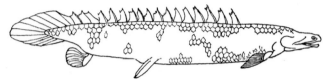

FIG. 68. Polypterus

fringe-like fin surrounds the tail. There is an air-bladder of two lobes which connect with the throat by a single duct, a close resemblance to lungs and performing the same functions. As in the Dipnoi the young have external gills.

HOLOSTEI

In this order are two American freshwater fishes of ancient lineage, Amia, the bow-fin, the Lepidosteus (Fig. 69), the gar-pike. To the primitive features of a spiral valve and optic

FIG. 69. Lepidosteus, the gar-pike

chiasma they have added the advanced possession of vertebrae with biconcave centra joined to one another by a ball and socket device. The fishes of both these genera are unfit for food and are, themselves, voracious feeders and great enemies of fishermen.

TELEOSTEI

All highly evolved fishes belong to this large order, the general anatomical features of which have been described above. At least thirty orders have been distinguished, many of which contain genera that are unfamiliar and relatively speaking are unimportant in that they illustrate no new structures, no significant advances in evolution. A selection of these orders may be mentioned and a few of their members named.

Isospondyli:	Clupea, herring; Sardina, sardine; Megalops, tarpon; Salmo, salmon
Haplomi:	Esox, pike
Ostariophysi:	Rutilus, roach; Cyprinus, carp
Apodes:	Anguilla, eel; Conger
Synentognathi:	Exocoetus, flying fish
Microcyprini:	Lebistes, guppy
Solenichthyes:	Hippocampus, sea-horse
Anacanthini:	Gadus, cod; Merluccius, hake
Percomorpha:	Perca, perch; Scomber, mackerel; Thunnus, tunny; Gobius, goby; Blennius, blenny; Sphyraena, barracuda
Heterosomata:	Limada, flounder; Solea, sole
Discocephali:	Remora
Plectognathi:	Mola, sunfish

CHAPTER TWENTY-NINE

AMPHIBIA

Craniates normally passing their adult life on land, but laying their eggs in the water, where their larvae lead an aquatic life until metamorphosis. The skin, usually moist and glandular, is permeable, so that no amphibian can live in salt water. The limbs are five-fingered. Lungs are present, yet much aeration of the blood takes place in cutaneous blood vessels. The heart has two auricles and one ventricle, as well as a truncus arteriosus and sinus venosus; there is little, if any, separation of arterial from venous blood. The kidney is mesonephric, and the ureter, genital ducts and rectum open at a common cloaca. The ear has a drum, no external pinna, and a single auditory ossicle: there are three semi-circular canals.

There are three orders:

ANURA (SALIENTIA) (frogs and toads)

Amphibia in which the adult has no tail and the hind limbs are large and may be used for jumping and swimming. The fore-limbs are short.

URODELA (newts and salamanders)

Amphibia with a normal and functional tail and limbs of approximately equal size. The body is often long. Gills may persist into the adult stage.

APODA (GYMNOPHIONA)

Amphibia in which there are neither limbs nor limb-girdles, and the tail is vestigial. Dermal scales are present. The body is adapted for burrowing.

The most striking characteristic of the Amphibia is that which is expressed in their name, since, with but few exceptions, an amphibian egg is laid in the water, hatches into an aquatic larva, and becomes at metamorphosis a normally terrestrial adult. The fish-like character of the larva and the four-limbed nature of the adult seem so obviously to suggest support for the hypothesis of recapitulation that, if for no other reason, the Amphibia deserve close consideration, justifying the popularity of the frog in all elementary courses of biology. Frogs, however, are not typical Amphibia: the newts and salamanders deserve at least equal attention. The appearance of a newt is familiar to most, and a significant feature is its tail, for a tail is a chordate characteristic and the tailless condition of the frog is a specialization.

The chief outward differences between the Amphibia and the fishes from which they have evolved are found in the nature of the skin and limbs. The pentadactyl limb makes its first appearance in the Amphibia. It proved itself to be one of the most useful and most adaptable of all the creations of evolution, and is one which, with its almost countless modifications, provides the pattern of the limbs of every higher vertebrate.

Its basic design is familiar. One long bone is followed by two long bones, at the end of which a group of nine small, irregular bones forms the wrist or ankle. Following these, five long bones support the palm of the hand or the sole of the foot, with 2, 3, 3, 3 and 3 short phalanges in fingers or toes. In the lower Amphibia the limb occurs in this typical form, with all 31 bones present and separate.

The skin, unprotected by dermal scales, is supplied with two different kinds of glands. There are mucous glands, which keep the skin moist; and others, producers of a milky liquid which causes intense vomiting if it is swallowed. When this unpleasant characteristic is associated with patches of vivid colour revealed when the limbs are moved, the protection against the amphibian's enemies is quickly established.

The moistness of the skin is essential because it is an important respiratory surface. Large veins lie just beneath it and are able to absorb oxygen directly from the air or from the water in which the animal is swimming. Although there are two lungs, their contribution to respiration is relatively small, and in many Amphibia they are reduced or even absent. This is specially true of the species that live in briskly running streams where the water is fully oxygenated.

The skin also has the ability to change its colour.

A further feature of the Amphibia, and evidence of their evolutionary advance, is their possession of vocal powers. The croaking of the common frog has been heard by almost everyone in spring time, and many species have very characteristic voices. The Natterjack toad says 'crac-crac-crac'; the bull frog calls for 'more rum, more rum'; and Aristophanes has immortalized the notes of the edible frog in the chorus line, 'Brek-ek-ek-ex, koax, koax'.

For the rest, the internal skeleton is largely bone, but cartilage is found in the skull and especially in the supra-scapulas.

The alimentary canal is fairly simple, the stomach appearing as little more than a broadening of the oesophagus. It is followed by a short duodenum and a coiled small intestine leading into a large intestine with no caecum. The rectum opens at a cloaca. In the loop between the stomach and duodenum is a compact pancreas, the duct from which is joined by the bile duct and the two enter the duodenum as a common hepato-pancreatic duct.

The heart has two auricles and one ventricle. A triangular sinus venosus opens into the right auricle and from the ventricle a conus arteriosus leads to three arterial arches on each side. These are the pulmo-cutaneous, the systemic and the carotid arches. The vascular system is almost wholly symmetrical.

The existence of two auricles and one ventricle causes venous and arterial blood to mix in the ventricle, and for many years it was the custom to seek for methods by which the two sorts of blood were more or less separated in the circulation. Present belief is that no such mechanisms exist and that mixed blood fills the whole of the system.

Large vessels, the pulmo-cutaneous arteries and their branches, carry blood to the lungs and skin for oxidation. The lungs are not the solid, spongy bodies that are found in mammals, but hollow sacs with branching capillaries forming much of their walls. The hollow, which should be filled by a steadily flowing tidal stream of air driven by the rise and fall of the muscles between the mandibles, has been exploited as an ideal place to live in by several species of Platyhelminthes. These are sometimes so numerous that they must displace a large fraction of the air which the lung ought to contain, and very considerably reduce its value as a respiratory organ. The skin, however, is a more than adequate alternative, and a dried frog has a poor chance of survival.

The kidneys are two compact red bodies attached to the

dorsal surface of the body cavity. They are mesonephric, and the testes, joined to them by the vasa efferentia, pass the spermatozoa through the kidney tubules and along the ureter to the cloaca. The ovaries occupy the same place, but the ova are too large to go through the kidneys. They are therefore shed into the coelom, enter the oviducts, and leave by the cloaca in the spring. There is thus a considerable difference between an amphibian dissected in the autumn, when the ovaries are large black and white objects filling most of the body, and an amphibian dissected in the spring, when the ovaries are indistinguishable and the ova are crowded into the lower ends of the oviducts.

The adrenal bodies are to be seen as small spots of tissue on the anterior part of the kidneys, and not, as in mammals, separated from them.

ANURA

The frogs and toads, which are probably the most familiar of the Amphibia, are also the most specialized. In many species the distinguishing feature is the increased length and strength of the hind legs, which enable frogs to swim readily and also to jump considerable distances on land. This habit is accompanied by several adaptations in the frog's body, such as the existence of two large veins bringing blood from the muscles of the hind legs and of a renal portal system which takes it first to the kidneys for purification before it enters the heart. The length of the ankle bones and of the toes also tends to increase the length of a leap, and the forward direction of the ilia, which unite with the last vertebra, provides a stable fulcrum on which to base the muscular effort.

Frogs and toads are widely distributed in all warm and temperate lands, where there are species that dig into the earth and a considerable number that climb into the trees. Some frogs are completely terrestrial, and the eggs, instead of being laid in water, are carried on the back or even in the mouths of one of the parents.

This, however, is about as far as Amphibia have progressed towards a complete emancipation from the water. Their continued dependence upon it, and their inability to withstand the salt water of the sea, are the chief limiting factors that have inspired the remark that while the Amphibia invaded the land the reptiles first occupied it.

URODELA

This order contains the typical Amphibia, often called salamanders if they are large, and newts if they are small (Fig. 70). They are lizard-like animals, living in cool, moist surroundings and hibernating during the winter. In spring they awake, travel to water, mate and lay their eggs. The males at this season are distinguished by the appearance of a dorsal crest.

One of the most remarkable members of the order is the

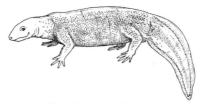

Fig. 70. A female newt

American Ambystoma tigrinum. The larva of this species was for long known under the name of the Axolotl, occurring especially in the lakes near Mexico City, where, still possessing gills, they reproduced themselves. In 1864 several specimens were sent to Paris, and here, after a few generations, it was found that metamorphosis could occur and turn the axolotl into an ambystoma. The ability to breed when in the immature larval state is uncommon, and the phenomenon is known as paedogenesis or neoteny. It occurs in a few other Amphibia.

APODA

This is a small order, surprising because its members have adopted a life underground. Imagine a small, thin newt with pointed skull and sharp teeth yet deprived of its limbs and tail, and the picture is almost exactly that of a Coecilian, and also almost that of an earthworm. The habits in this group, so far as they are known, are most unusual. Their home is a burrow connected to a pond or a stream, and at its far end the female lays the eggs and lies curled round them. They hatch after losing their non-functional gills into larvae with internal gills, and later metamorphose into adults with lungs. Their relationship to the extinct Stegocephalia is shown by the bones of their skulls and the presence of very small scales in the skin.

The evolutionary status of the Amphibia gives them an interest which compensates for their relative unsuccess in the world of animals. They seem to be obviously derived from the Rhipidistian fishes mentioned in the last chapter, and they must also be regarded as the ancestors of the reptiles, which made a greater success of life on land. In this connection mention should be made of one of the world's most interesting extinct animals, Seymouria, which holds almost equally the balance between an amphibian with early reptilian characters and a reptile with traces of its amphibian ancestry. These creatures have come from the Carboniferous and Permian strata, Seymouria itself being named after the town in Texas where it was discovered. In appearance it was probably like a lizard with short legs, a slower mover than the true lizards of today. Its skeleton unites amphibian and reptilian characters in a peculiar manner.

When one looks at the chordate phylum as a whole, one should be forcibly struck by the great theoretical importance of a few genera which seem to be more usually neglected than they deserve. The extinct Rhipidistian fishes have features which indicate clearly that the class of Amphibia was evolved from fish-like ancestors closely allied to such genera as Osteolepis. This is followed by Seymouria, which forms a link between the Amphibia and the earliest Reptilia. Proceeding, we meet Archaeopteryx, the well-known name of an animal that can only be described as a reptile-like bird or a bird-like reptile. Lastly Cynognathus is a clear example of a reptile beginning to acquire the characteristics of a mammal.

Popular speech often refers to 'the missing link', but here there are links which are certainly not missing, save in the sense that they are no longer to be found alive in the world. But their one-time existence as forms intermediate between the definitive classes of vertebrates is an almost incontrovertible proof that organic evolution has, in fact, taken place.

CHAPTER THIRTY

REPTILIA

Craniates that are completely terrestrial throughout their lives, showing extensive adaptive
radiation. The exoskeleton consists of horny
scales and also bony plates. The heart has four
chambers and respiration is by means of lungs
only. The kidney is metanephric and there is a
cloaca. There are twelve pairs of cranial nerves
and a single auditory ossicle. The eggs are large
and yolk-filled; they are laid in a shell and the
embryo is provided with an allantois and protected by a fluid-filled amnion. Reptiles are
usually cold-blooded and do not penetrate the
polar regions.

There are four orders:

RHYNCHOCEPHALIA

Rare, lizard-like reptiles in which the persisting pineal eye
penetrates the skull and skin.

CHELONIA (TESTUDINES) (tortoises)

Reptiles in which the body is enclosed by dorsal and ventral
shields of bone covered with horny, epidermal scales. There are
no teeth and there is no sternum.

CROCODILIA (LORICATA)

Large reptiles adapted to life in fresh water. The tail is
laterally compressed for swimming, and the jaws are furnished
with many sharp teeth.

SQUAMATA

Terrestrial reptiles in which the body is covered with scales.
The jugal and squamosal bones do not meet.

SUB-ORDER LACERTILIA (lizards)

Squamata with long tails, with the rami of the lower jaw
united and with limb-girdles functional or vestigial.

Squamata with long bodies, widely expansible mouth with a sensory tongue and often poison fangs. There are neither limbs nor limb-girdles.

The class is sometimes divided into three sub-classes, Anapsida, Synapsida and Diapsida.

The first truly terrestrial vertebrates, the reptiles owed their advance to a new type of embryonic development. Like their ancestors, the Amphibia, they lay eggs, but the ova are fertilized in the maternal oviducts and the embryo is fully protected. The obvious and external form of this protection is the

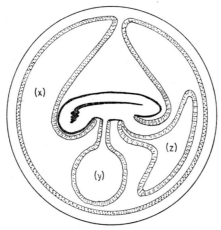

Fig. 71. Foetal membranes. (x) Amnion. (y) Yolk sac. (z) Allantois

egg shell, which, though not as hard as the calcareous shell of a bird's egg, is a tough, resistant covering.

Inside the shell the developing reptile is associated with the so-called embryonic membranes, the yolk sac, the allantois, and the amnion (Fig. 71). The yolk sac, as its name implies, contains a quantity of nutrient protein and is connected by a duct to the alimentary canal of the embryo. The allantois, an outgrowth of the hind end of the gut, is a hollow structure which may become large enough to surround both the embryo and its yolk sac, but always a portion of it is close to the porous shell through which it receives oxygen. It is, therefore, the means by which respiration is effected, and it may also serve the valuable function of a receptacle for excreta, which accumulate within it. In higher chordates, the allantois becomes the placenta, by which the nutrition of the embryo is achieved.

The amnion is formed as a result of the movement of the embryo towards the centre of the egg: thus it takes the form of a double protective envelope around the embryo. The space between the two layers is filled with liquid, the amniotic fluid which acts not only as a protection against shock but also as a preventive of desiccation. In consequence, the eggs can be laid and development can proceed on dry land, which in turn leads to the omission of a larval stage.

The internal organ systems show a general progress from the amphibian condition, with residual signs of their origin. The alimentary canal, for example, ends like that of an amphibian at a cloaca.

The heart is completely, or almost completely, divided into right and left halves, so that a double circulation, like that of a human being, is established. This implies the sole use of lungs as respiratory organs, for the temporary visceral clefts never carry functioning gills. In many reptiles a more or less complete palate carried the inspired air to the region of the trachea at the back of the mouth, this being a mammalian rather than an amphibian feature, but the pattern of the blood vessels is virtually symmetrical, and the aortic arches originate directly from the ventricle.

The brain is relatively larger than that of an amphibian, and like the brains of birds and mammals carries twelve pairs of cranial nerves. The ear has no external part; it contains only one auditory ossicle, but has an elementary form of cochlea.

The excretory organs are kidneys, which are of the metanephric type, like those of a mammal and unlike those of an amphibian. The chief features of a metanephric kidney are its dissociation from the vasa deferentia of the male genitalia and its function in water economy, resulting in the excreta leaving the reptilian cloaca consisting more of uric acid and less of urea than in Amphibia.

The gonads occupy permanent positions on the dorsal wall of the body cavity. The vasa deferentia are joined by the ureters before reaching the cloaca, but the oviducts open directly into it. There is sometimes a pair of eversible sacs which can act as intromittent organs, a paired condition which persists into the Prototheria.

The Tuatara, Sphenodon punctatum, is one of the world's important animals. It is found only in the region of Cook's Strait between the north and south islands of New Zealand, and its general appearance is that of a small crocodile with a head rather too big for its body (Fig. 72). It is covered with brown granular scales and has a tail with a saw-like upper edge, but above and beyond all this is its amazing possession of a third,

FIG. 72. Sphenodon punctatum, the tuatara

median, eye in the middle of its forehead. In other reptiles this is present only as a lobe in the brain, in the fishes and amphibians it is more conspicuous as the pineal body.

There is no other animal alive in which the pineal body passes through the skull and carries an apparently functional eye at its extremity. It has the normal structure of a small and rather primitive eye, typical of the chordate phylum. Most probably it was more efficient and more valuable in some earlier reptiles than it is in Sphenodon today, so that Sphenodon is one of the rather large number of animals popularly described as living fossils.

The Tuatara is a slow-moving, lazy reptile, living in short burrows which it digs itself. Advantage is taken of the shelter thus provided by petrels living in the neighbourhood, a very unusual partnership between a bird and a reptile in a shared home.

The reptile's eggs are laid a few inches underground and are hatched by the warmth of the sun with no attention from their parents. The young develop uncommonly slowly and usually do not emerge for more than a year after they have been laid. They are independent from the beginning of their lives.

Because of the attraction of its unique character Sphenodon was at one time threatened with extinction. It is now protected by law and its survival should be assured.

CHELONIA

Tortoises and turtles are probably more heavily armoured than any other animals, and the rigid casing protecting them is familiar to all. It is composed of two elements, an upper domed carapace and a flattened ventral plastron. Each consists of bony plates covered with sheets of horn, generally called tortoise-shell. The carapace is derived from neural spines and ribs flattened almost beyond recognition, the plastron from clavicle and interclavicle, but the sternum has disappeared. Almost equally unexpected is the absence of teeth; a layer of horn covers the jaws and forms a pair of blades with which the food can be cut.

The animal can withdraw itself almost completely within its shell. There are two ways of protecting the head. Among the more primitive genera of the southern hemisphere it is simply turned to one side towards the shoulders and under the fore-edge of the carapace. Others draw in their head by a bending of the flexible neck into an S shape.

The Chelonia have produced both terrestrial and aquatic forms, the latter being commonly distinguished as turtles. These live either in marshes or in the sea, and marine turtles have in general lost some of the weight of their shells and have converted their legs into paddles, with web between the toes.

The most remarkable feature of land tortoises is the great age to which they may live and the size they may attain. The giant tortoises of the Galapagos and other islands may reach a length of four feet and a weight of a quarter of a ton, while a tortoise taken in the Seychelles in 1766 lived until 1918. The most probable explanation of this phenomenon is a tendency, general among reptiles, to continue growth without a limiting factor, combined with an absence of natural enemies in their isolated environment.

CROCODILIA

The crocodiles and alligators form a comparatively uniform group in which adaptation to an aquatic life is strongly marked, so that they do not occur in varied environments. They are normally slow movers, with short legs and heavy bodies, and much of their time is spent lying in the water or occasionally basking in the sunshine on the bank.

Their tails are laterally flattened, they are reasonably good swimmers, and a well-known adaptation to the mode of life is the raising of their eyes and nostrils above the level of their heads. This enables them to remain partially hidden under water and at the same time to see and to breathe. Thus they can seize any bird or other animal unfortunate enough to approach them, while at the same time they gain benefit from a palate which provides passage for air from the nostrils to the neighbourhood of the trachea.

The hind legs of crocodiles are conspicuously longer than the fore legs, a feature of some of the earliest reptiles which appeared, but most of these are now extinct while the crocodiles remain, surviving by virtue of their specialization for life in fresh water.

The distinction between crocodiles and alligators is not outwardly conspicuous. The difference most easily to be seen concerns the fourth tooth of the lower jaw. In alligators this large tooth fits into a pit in the upper jaw; in crocodiles it merely rests in a notch when the mouth is closed.

Crocodiles seem to foreshadow mammals in having a heart in which the ventricle is completely divided, as well as a rudimentary form of muscular diaphragm inside the body cavity. But they lay hard-shelled eggs, from twenty to ninety at a time.

LACERTILIA

The lizards form a large group of over two thousand species, with so wide a distribution in all except the coldest regions, and so many adaptations to different modes of life, that they must be considered the most ambitious and most successful of living reptiles. In addition to the rather shy species of this country there are species that burrow or swim or fly; and there are lizards in forests, in deserts and on mountains. One group, typified by the slow-worm, has no external traces of limbs and so even more closely resemble the snakes to which lizards are nearly related.

The body is covered with small horny scales that are shed at intervals and which sometimes overlie bony plates. The legs are rather thin, but enable many lizards to move rapidly. The toes may be webbed, or provided with adhesive pads, or otherwise modified for use on vertical surfaces or in soft sand. A very characteristic feature of lizards is the long tail in which the vertebrae have 'cleavage planes' where they can easily break. This enables the lizard to escape from the grasp of a pursuer, after which a new tail is regenerated from the stump.

Lizards lay eggs, but among some families the eggs are retained until the young are ready to hatch immediately, i.e. they are ovoviviparous.

There are many families of lizards, of which by far the most distinctive is that of the chameleons of Africa and central Asia. Their ability to change their colour has brought their name into ordinary speech, yet it is a power which they share with fishes, amphibians and other reptiles, some of which surpass them in this respect. The change is due to the presence in the dermal cells of pigment granules which can move about within the cell.

Fig. 73. The slow-worm or blind-worm

Various stimuli provoke the change, light, warmth and even emotion among them. Zoologically a more remarkable feature is the independent movement of the eyes, which can be turned through wide angles and often stare in different directions.

The geckos are small lizards of wide distribution which announce their identity by a clicking chirp sounding very like the word 'gecko'. Many lizards have no voice. The skinks are another world-wide family, some of which approach two feet in length, while others are limbless.

The real limbless lizards are, however, the slow-worms of the family Anguidae. The familiar English species (Fig. 73) is often given the name of blind-worm, quite inappropriately as it has good eyes. It is an attractive silver-grey species with no vestige of legs and it lives almost solely on slugs. It is one of the ovoviviparous kind, the young escaping from the egg-membrane almost as soon as they are laid.

By far the most dramatic of the lizards are the monitors or

dragons belonging to the family Varanidae. They are large and apparently ferocious animals, found in South Africa, Asia and Australia, where they are often feared and believed to inflict venomous bites. In fact, they have no venom, and some species use their long lashing tails as their chief weapons. The giant of the family is the Komodo dragon, Varanus komodoensis, which reaches a length of ten feet and is a much more heavily built creature than other members of the family. In its native island of Komodo in the East Indies its chief food is small pigs.

OPHIDIA

Snakes are as numerous as lizards but are far more important to man. Anatomically their chief characteristics are the absence of limbs, the forked tongue which is sensitive to odours, the large broad scales on the lower surface of the body, and the curved backward-pointing teeth.

Many snakes are venomous. The venom is secreted by modified salivary glands and is conveyed by a duct to the base of the fangs which are at the front or the back of the upper jaw. The sharp fangs of snakes, however, do not bite their prey, which is usually swallowed whole, and, as is well known, may be larger in size than the snake's normal diameter. To allow such massive quantities to pass into the oesophagus, snakes have the unique power of dislocating the bones of the lower jaw and also moving them apart, so that the gape of the mouth is enormously stretched. Meanwhile, the glottis at the top of the trachea is pushed forward between the mandibles and breathing can continue. The digestion of a single meal may be prolonged.

The smooth, gliding movement of snakes is the result of a characteristic relation between the ribs and the scales of the belly. The ribs can be moved backwards and forwards, and as they go backwards the edge of the attached scale meets any projection on the surface of the ground and propels the snake.

Three families of snakes, the Typhlopidae, Leptotyphlopidae and Ilyssidae, are harmless burrowers which make no appeal to popular imagination. On the other hand, the Boidae, containing the boas and pythons, are remarkable for their great size and may attain twenty-five or perhaps thirty feet in length. These snakes are not venomous: they kill their prey by coiling themselves round the body of the victim and squeezing it to death. They often climb trees and are fond of lying in water.

Pythons lay eggs to the number of about a hundred, encircle

the large clutch and, rising in temperature, actually incubate them. Boas usually produce their young alive.

Most snakes belong to the large family Colubridae, very widely distributed in the tropical and temperate zones. The English grass snake is a harmless member of this family and is a good swimmer, but most of the species are arboreal. The cobras of Africa and Asia, famous for the hood made by extending the ribs of the neck region as a warning or threatening reaction, belong to this family, as do the much dreaded Hamadryad of India and the Mamba of Africa.

The family Viperidae contains the vipers and rattlesnakes, all of which are venomous. The vipers and puff-adders are generally terrestrial and viviparous, and the flattened head of the familiar British adder is a characteristic of the family. The rattlesnakes, Crotalinae, show a wider choice of environment and species may be found in trees, in water, as well as underground, chiefly in America. The rattle is derived from a terminal scale which is not shed when the snake sloughs its skin but remains covering the tip. The noise is produced by shaking the tail.

The living reptiles briefly described above are the survivors of a far more impressive group which formed the dominant Chordata of the Mesozoic Era. During this time they achieved a remarkable degree of adaptive radiation, evolved some species which grew to an almost incredible size, and declined suddenly, though not before they had given rise to the earliest birds and mammals.

Their radiation is shown by the fossil remains of reptiles which evidently swam or flew. In the seas were the animals now known as Ichthyosaurs and Plesiosaurs. The former had long jaws with a large number of teeth; their limbs were paddles and the tail was a vertical fin. The latter had a small head and a long neck, the fore-limbs were much larger than the hind-limbs and the short tail had no fin.

The flying reptiles were the Pterodactyls. Like birds, they flew by means of wings which were modified fore-limbs: the fifth digit was lost and the fourth, greatly enlarged, supported a membranous wing.

The Dinosaurs, which capture the imagination because of their size, were of two kinds, distinguishable as the reptile type and the bird type, or the Saurischia and Ornithischia respectively. The former included both carnivores like Brontosaurus, and herbivores like Diplodocus, animals which may have weighed as much as thirty tons. The latter had limbs and limb-girdles which resembled those of a bird, and their hind legs were much the bigger pair. Iguanodon is probably the most familiar example: it walked on its hind legs and reached a height of fifteen feet: Tyrannosaurus was nearer fifty feet long and is reckoned to have been the largest land animal that has ever existed, for it was also about twenty feet high.

In the same sub-class, Synapsida, are to be found the fore-runners of the mammals, often called the Theromorpha. The earliest forms of these belong to the Permian and probably

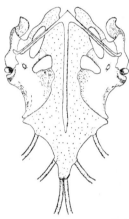

FIG. 74. Reptilian pectoral girdle

arose from ancestors related to Seymouria. The most far-reaching of the essential changes was the rotation of the limbs, which, instead of being almost horizontal, gradually came to be more and more vertical and so to raise the trunk above the ground. This made movement easier and in turn led to greater speed; or, in other words, crawling gave place to running.

Certain other Theromorph characters are interesting and important. The establishment of a palate between air and food passages made it possible to retain food in the mouth for more thorough chewing, in correlation with which a differentiation of teeth began to appear.

The pectoral girdle (Fig. 74) included a coracoid and pre-coracoid, a duplication found in the marsupials. In the skull the auditory capsule dropped and facilitated the expansion of the cranium to accommodate the growing cerebral hemispheres. In some genera the lower jaw articulated with the squamosal as in mammals today, and the quadrate became the incus of the ear.

These steps towards the Mammalia are well illustrated by one of the most interesting of extinct animals, Cynognathus (Fig. 75). This 'cross between a lizard and a dog' had reptilian

FIG. 75. Cynognathus

skull and jaw support, and the digits of its limbs were those of a reptile. On the other hand, it had a dog-like muzzle with the cusped molars of a carnivore and a dentition in general more like that of a mammal than a crocodile. The skull articulated with the atlas by two condyles and the vertebrae were to some extent differentiated. It is unfortunate that fossils have not yet traced the transition to milk, warm blood and hair, but bone remains indicate that these changes took place during the Triassic epoch, and that both reptiles and mammals were prosperous land animals during the succeeding Jurassic period.

CHAPTER THIRTY-ONE

AVES

Warm-blooded craniates, clearly related to the reptiles, covered with feathers and capable of sustained and directed flight by means of wings which are modified pectoral limbs. The hind limbs are scale-covered and are strengthened by a degree of bone fusion. The sternum is large and has a keel to which the pectoral muscles are attached. There are no teeth, and the jaws are covered by horny sheaths, forming the beak. The heart is four-chambered and the right aorta alone persists. The kidney is metanephric and there is a cloaca. Internal air-sacs contain a large volume of reserve air. The eggs are large, protected by a calcareous shell, laid in a nest and incubated with conspicuous parental care.

There are two sub-classes:

ARCHAEORNITHES

Extinct birds, with teeth and a jointed tail.

NEORNITHES

Modern birds without teeth.

The characteristic feature of birds is their covering of feathers, anatomically speaking their most conspicuous distinction from the reptiles from which they have evolved. They also owe at least part of their success to internal improvements. They are warm-blooded, and their rapid metabolism is shown both by their powerful flight in the air and their restless activity on the ground. Other external features are the horn-covered beak, the shortness of the tail, and the legs which are covered with reptilian epidermal scales and end in four toes, one of which is pointed backwards.

Feathers, which have neither parallel nor imitation in any other group of animals, are of five kinds: remiges or quills on the wings, contour feathers, coverts, filoplumes and down. Structurally all these are similar, with a central rachis which in quills is hollow at the base. The rachis bears the vane, which on each side consists of flat plates, the barbs. These are joined together by small hooks, or barbules and barbicels which weld the vane into a continuous elastic surface. In the contour feathers, which cover almost the whole body, the barbs are more easily separated than in quills; and the bases of both these types of feather are covered by the small coverts. Filoplumes have a bare rachis with a tuft of barbs at the top, and down feathers are of barbs only, with no rachis.

The alimentary canal of a bird begins at its bill or beak, and the many different shapes which the beaks assume are often illustrated as examples of the adaptation of structure to

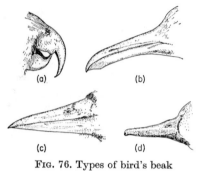

FIG. 76. Types of bird's beak
(a) Parrot.   (b) Gull.
(c) Rook.   (d) Duck.

function (Fig. 76). There is here little need to do more than to remind readers of such beaks as those of the eagle, the duck and the woodpecker.

The oesophagus leads to the crop, a place of temporary storage, followed by the stomach and in most birds by the gizzard. This is an organ of considerable muscular strength, which takes the place of teeth as means of triturating the food. It contains a permanent supply of pebbles which, acting like millstones, provide an unexpected parallel to the bits of grit in the gizzard of an earthworm.

Two bile ducts enter the duodenum: the intestine is long and leads to a short rectum and cloaca.

The heart, like that of a mammal, is four-chambered, but from the left ventricle only the right aortic arch has persisted. From this arise two innominate arteries, which give off the carotids. The venous system includes three venae cavae and a small renal portal vein. The respiratory organs are a pair of lungs, which are in communication with an elaborate system of air sacs. These are the chief internal peculiarities of birds. The trachea divides into two bronchi, and the syrinx is situated at the point of division. It contains a semilunar membrane which produces the notes of a bird's song and is therefore analogous to the larynx at the top of a mammalian trachea. The air sacs lie among the viscera and are even to be found in the bones. They probably serve as a reservoir for an extra supply of oxygen during rapid or sustained flight.

The kidneys are two conspicuous bodies on the dorsal wall of the coelom. They are metanephric. Each is of three lobes, from the posterior of which a ureter runs to the cloaca. There is no bladder, and the excreta contain uric acid, not urea. Some water is re-absorbed by the rectum.

There is only one ovary, a condition which recalls that found in the dogfish and is also doubtless to be attributed to the large size of the ova. There are two testes whose ducts deliver the sperm to the cloaca: there is no intromittent organ.

The brain has large cerebral hemispheres, witnesses of the intelligence and general educability of birds. There are twelve pairs of cranial nerves and by far the most efficient sense organs are the eyes.

The large size of the eyes makes the skull of a bird recognizable at a glance, and a second feature is the almost complete fusion of the bones so that their separate individuality can be determined only in the young chick. The quadrate helps to support the lower jaw, which, with the upper jaw, forms the beak. There is one columella auris and one occipital condyle.

There are from 13 to 15 cervical vertebrae, which after the second carry short cervical ribs. Six lumbar, two sacral and five caudal vertebrae are all fused to make one long sacrum (Fig. 77) to which the pelvic girdle is united.

The sternum is a broad plate with a large central keel to which the wing muscles are attached, and the pectoral girdle consists of blade-like scapulae, strong coracoids and united clavicles, very familiar as the merry-thought. Much the most interesting part of a bird's skeleton is found in its fore-limbs, which provide an example of extreme specialization of the pentadactyl limb. The humerus is rather short, the radius and ulna are well separated and are followed by two carpalia, the radiale and ulnare. Three bones of the distal row of carpals are united and all are fused together to the first, second and third metacarpals to form a unique complex, the carpometacarpus.

Fig. 77. Pelvic girdle of a seagull

There are traces of the first three digits, consisting respectively of 1, 2 and 1 phalanges. The relation between these bones and that of the human hand is shown at Fig. 78.

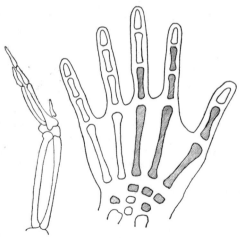

Fig. 78. Bones of a bird's wing and a human hand. (The shaded human bones are found in the bird's wing.)

214

In the hind-limb there is a comparable tarsometatarsus, with the additional feature that one toe is turned backwards.

ARCHAEORNITHES

Although this sub-class contains no living species it is too important to be neglected. Birds have left a comparatively small number of fossils, a fact that is attributed in part to the lightness of their bones, so that Archaeopteryx lithographica, of which a feather and two skeletons were found in the Ottmann Quarry at Pattenheim, Bavaria, in 1861 and 1877, is one of the most famous fossils in the world.

The popular description of Archaeopteryx as a bird with teeth is perhaps justified because it possessed feathers, but its truly avian characters were not much more than an opposable toe, a 'wish-bone' and backward direction of its pubic bones. Against these four features must be set the fact that it had a long tail of 20 vertebrae, a large fore-limb, claws on all its fingers, teeth in its jaws and a reptile's brain. Its sternum had no keel, which suggests that it did not fly so much as glide or parachute. It is, however, definite proof that a passage was made from reptiles to birds, and that the first birds must have belonged to the Jurassic epoch. The true fliers were scarcely to be seen before the Cretaceous.

NEORNITHES

About eight thousand species of birds belong to this group. The idea of dividing them into the Ratitae or flightless birds and the Carinatae for the rest, has been abandoned because the former are not more closely related to one another than are the birds of the other orders. In consequence, the systematics may be written thus:

Super-order Odontognathae
    Order Hesperornithiformes
    Order Ischthyornithiformes
Super-order Impennae
    Order Spheniscformes   penguins
Super-order Neognathae
    Order Struthioniformes   ostriches
    Order Rheiformes   rheas
    Order Casuariiformes   emus, cassowaries

Order Colymbiformes	divers
Order Procellariiformes	petrels
Order Pelecaniformes	gannets
Order Circoniiformes	herons
Order Anseriformes	ducks
Order Falconiformes	eagles
Order Galliformes	game birds
Order Ralliformes	rails
Order Charadriiformes	cranes
Order Columbiformes	pigeons
Order Cuculiformes	cuckoos
Order Psittaciformes	parrots
Order Strigiformes	owls
Order Caprimulgiformes	nightjars
Order Coraciiformes	kingfishers
Order Apodiformes	swifts
Order Piciformes	woodpeckers
Order Passeriformes	finches

Whenever a large number of species are to be arranged according to the principles of zoological classification difficulties are always encountered, because the richness of the material shows itself in the existence of an unwelcome proportion of intermediate forms. This is particularly true of the class Aves, where many zoologists would be inclined to say that the accepted system was practically useful rather than truly phylogenetic. In addition to this, birds of all kinds show an unusual similarity of structure, and uniformity is not a contribution to a system which is essentially based on differences.

There is another unusual feature about the systematics of birds. In most other phyla and classes an arrangement of the orders and families in what appears to be an evolutionary sequence shows a gradual increase in body size. Among birds the reverse is true, and the tendency of evolution is towards smaller and smaller bodies.

Apart from this, the two main problems which birds present to the zoologist are the origin of flight and the habit of migration. Neither of these can yet be said to have found a perfect explanation.

An ability to fly has been acquired by insects and by three different groups of vertebrates, the reptiles, birds and mammals. The flying fishes may perhaps be considered at the same time.

When fishes 'fly' the propulsive force originates from the swimming movements of the tail, which drives the fish out of the water in a manner no different from the leaping of a salmon over a weir. The enlarged pectoral fins are held at right angles to the body and although they scarcely move they no doubt prolong the time spent in flight.

Naturalists have described flying frogs, flying lizards and flying snakes from many parts of the world. All these animals are gliders rather than fliers. The frogs have long fingers and toes and the webs make gliding a possibility: the lizards, like Draco, have lateral skin folds supported by extensions to the ribs. The snakes are really parachutists, using a concavity of their lower surface made by expanding the ribs and retracting the muscles below them.

True flight has been achieved by pterodactyls, bats and birds. The description of the origin of a bird's flight at one time varied between an extension of running jumps, that is to say an improvement on the fishes' method, or an elaboration of the gliding of the reptiles. When the sternum of Archaeopteryx was detected and found to have no keel, which is an essential for wing muscles, the gliding hypothesis was obviously favoured.

The seasonal migration of birds, in spite of years of fact-finding, experimenting and theorizing, remains more difficult to explain even than the origin of flight.

CHAPTER THIRTY-TWO

MAMMALIA

Craniates, normally terrestrial, showing a high degree of structural, physiological, and behavioural adaptation to all types of environment. The skin produces hair, and modified skin glands produce milk for the early nourishment of the young. In the brain the cerebral hemispheres are very large and the optic lobes are divided. The heart has four chambers and there is a complete separation of arterial and venous blood: only the left aortic arch persists. The ear has an external pinna and three auditory ossicles. A muscular diaphragm separates the thoracic from the abdominal portions of the body cavity. There are seven cervical vertebrae, and inter-vertebral discs exist between the centra. The kidney is metanephric, and anus and urino-genital apertures are distinct.

There are three sub-classes:

PROTOTHERIA (monotremata)

Reptile-like mammals which lay yolk-containing eggs and nourish their young on milk secreted by modified sudorific glands. The brain is comparatively small. There is a cloaca. Separate coracoid and precoracoid are present in the pectoral girdle, and there is also an interclavicle.

METATHERIA (marsupialia)

Mammals which are viviparous and which carry their young in an abdominal pouch, feeding them on milk from modified sebaceous glands.

EUTHERIA

Mammals in which the young are born in an advanced, almost independent condition after a long gestation. The brain is highly developed. The sub-class contains all the most familiar and domesticated four-footed beasts. (See Chapter Thirty-three.)

PROTOTHERIA

Only three genera represent the duckbills, the most primitive group of mammals alive today. They are:

Ornithorhynchus, with O. anatina, the duck mole
Tachyglossus, with T. aculeatus
Zaglossus, with Z. bartoni.

The first two of these are Australian, the last is found in New Guinea, and all three show obvious remnants of a reptilian ancestry. As mammals they have hair, a diaphragm and the ability to produce milk; and they combine these features with such reptilian characters as large clavicles with an inter-clavicle between them, epipubic bones (Fig. 79), small brains

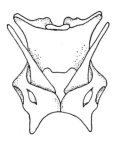

FIG. 79. Pelvic girdle of Tachyglossus

and a cloaca. Most conspicuously, they lay eggs and the young that hatch from them are fed on milk. This is a fluid secreted by modified sudoriferous glands in contrast to the modified sebaceous glands that produce the milk of higher mammals, and it is lapped up by the young from depressions in the mother's abdomen.

With all this, specialized features exist. The duck mole, often called the duck-billed platypus, lives in or near rivers and swims easily. It digs burrows in the banks, at the far end of which, twenty feet away or more, a rounded nest-chamber is made and lined with leaves and grass. Here the female lays one, two or three eggs. The male duck mole has an unusual weapon in the form of a thorny spur arising from the inner side of each hind leg and connected with a venom gland.

The family of Echidnidae, which contains the other two genera known as spiny ant-eaters, is represented by several species and spreads into Tasmania and New Guinea. They are burrowing animals whose diet is chiefly ants. The egg is carried about in a pouch where it hatches and where the young one remains for some weeks in a manner reminiscent of the characteristic of the true marsupials.

METATHERIA

These primitive mammals are now confined to Australia, New Guinea and South America, where their survival is usually attributed to the absence of effective eutherian competition.

Their best-known characteristic is their habit of carrying their young in a marsupium or abdominal pouch, and the significant fact is that the young are born when they are very small and undeveloped. A six-foot kangaroo produces an infant scarcely an inch long. This can find its way along a salivated track made by the mother to the pouch, where it seizes a nipple. It remains attached to this continuously for some time, while milk is not sucked but is forced or pumped into it. A remarkable prolongation of the larynx as far as the nose prevents the baby from choking, and is so unlike any organ in any other mammal that it justifies the description of the newly-born kangaroo as a larva.

These features are specializations. The primitive nature of the sub-class is shown by its anatomy.

The skull has no tympanic bulla and has an incompletely ossified palate, points which it shares with the skulls of the Insectivora. The lacrymal duct opens outside the orbit; there is no placenta; there is a cloaca.

Formerly of world-wide distribution, yet for long confined to the regions mentioned above, the marsupials have shown within their sphere a degree of evolution which forms a fair parallel to that of the Eutheria. There are carnivorous genera, such as Sarcophilus, the Tasmanian devil, which have pointed canine and cusped molar teeth; there are herbivorous genera, of which the kangaroos and wallabies are representatives, with ridged molars; and there are marsupials that live in trees, others that burrow like moles; there are cat-like, mouse-like marsupials, rabbit-like, mole-like marsupials, blind underground tunnellers with horn-protected noses and two large triangular claws on their forefeet, and otter-like marsupials which live in fresh water.

In the New World the chief marsupials are the opossums of the family Didelphidae. These are nocturnal animals living in the trees, and the first digit of each foot is opposable, helping to grasp the branches. They have ten upper and eight lower incisor teeth, and a long, scaly prehensile tail. An opossum may be as large as a cat or as small as a mouse and is remarkable for its habit of aestivation or undergoing long periods of sleep in the summer.

CHAPTER THIRTY-THREE

EUTHERIA

Mammalia in which the young are retained in the uterus until they are well-developed and closely resemble the adults. The placenta is fully elaborated. Mammary glands are characteristic.

The Eutheria comprise all the most highly evolved mammals, including Man. They may be conveniently divided into four cohorts or super-orders: thirteen of whose seventeen orders will be mentioned:

UNGUICULATA

Eutheria of primitive nature, with nails or claws on the digits, simple brain, incomplete palate and persistent clavicle and centrale. Insectivora, Chiroptera, Primates, Edentata.

GLIRES

Eutheria showing the primitive features of the Unguiculata, but with teeth and jaws adapted for gnawing: to this is added an extreme fertility expressed by the production of several large litters a year. Lagomorpha and Rodentia.

MUTICA

Eutheria adapted to a marine life and with a consequent fish-like shape. The tail takes the form of two horizontal flukes, the fore-limbs function as stabilizers, the hind-limbs are vestigial. Cetacea.

FERRUNGULATA

Eutheria with hooves or claws, with the distal bones of the limbs more or less fused and the digits often reduced in number. The alimentary canal is elaborated. The group is the most highly evolved of all the Chordata—Sirenia, Carnivora, Pinnipedia, Proboscoida, Artiodactyla, Perissodactyla.

INSECTIVORA

Typical insect-eaters are mostly small animals, with sharply pointed teeth and an unexpected tendency to include spines more or less freely among the hairs of the coat. Their feet are provided with claws, their pectoral girdle has retained the clavicle, and the cerebral hemispheres are quite smooth. The testes, sometimes near the kidneys, never descend to a scrotum. All these characteristics are in accord with the primitive nature of the order; they recall the Reptilia, and thus indicate its relation to the Chiroptera and the Lemuroidea of the Primates. A further significant fact is their absence from Australia and comparative rarity in South America.

The more important sub-order contains the familiar shrews, moles and hedgehogs.

The shrews or Soricidae are a large family which includes some of the smallest living mammals; the pigmy shrew, Sorex minutus, for example, is about 5 cm long.

CHIROPTERA

Bats are the only mammals that have succeeded in developing the power of flight, the specialization that chiefly separates them from the Insectivora to which they are related. The wing is one of the many modifications of the pentadactyl limb, and consists of an extension of the skin, sometimes called the patagium, stretched from the shoulder to the thumb, supported by the bones of the second to the fifth digits, and thence to the side of the body and hind legs. In some species it also fills the space between the hind leg and tail.

There is a strong pectoral girdle, but the pelvic girdle and hind leg are relatively small. As in birds the normal body temperature is high. Bats are nocturnal and their most surprising feature is their ability to avoid obstacles during flight. Many experiments have demonstrated this, and it is now known that it is a consequence of a system resembling the radar of aircraft. The bat in flight emits a high-frequency squeak, this is reflected from objects ahead, and the returning echo, picked up by the sensitive ears, enables the bat to direct its flight.

Thanks to their ability to fly, bats are found all over the world, even including isolated oceanic islands.

Bats are roughly divisible into large bats, Megachiroptera, normally fruit eaters with flat molar teeth; and small bats, Microchiroptera, insect eaters with cusped molars. The former, the largest of which has a wing span of about five feet, include the so-called flying foxes of Asia and Australia; the latter group spread into temperate regions and include the common pipistrelle, as well as the formidable American vampire, Desmodus rufus.

PRIMATES

Man, the most intensively studied of the Primates, cannot claim to be a member of the most highly organized order of mammals. He is, however, one of an interesting group, with many primitive morphological features relating it to the Insectivora. But these have found compensation, and more, in certain specializations that have resulted in conspicuous advances.

Primates are essentially adapted to an arboreal life, and one of their valuable adaptations is the forward direction of the lines of vision of their eyes. Supplementing this binocular vision is the opposable thumb, which enables the hand to do so much that would otherwise be impossible. Coordination of hand and eye has stimulated development of the brain, from which has proceeded a development of the cerebral hemispheres and their surrounding cranium, and with this the growth of intelligence, speech and creative thought.

This impressive advance has taken place in a body that still has a complete clavicle, shows no fusion between radius and ulna or between tibia and fibula, retains all five digits on both pairs of limbs.

The sub-orders include the Prosimii or Lemuroidea and the Simiae or Anthropoidea. In the former are the lemurs, with dog-like muzzles, the popular bush-baby, the aye-aye and the tarsier. The aye-aye of Madagascar has a long thin third finger with which it extracts insects from holes in trees: the tarsier from the East Indies has large eyes, but is chiefly of interest because it is believed to be a modern representative of the ancestors of man.

The Anthropoidea are the monkeys and apes. Catarrhinae, Old World monkeys, have narrow noses and a non-prehensile tail; Platyrrhinae, New World monkeys, have broad noses and prehensile tails.

The great apes are the gibbons, orang-utans, gorillas and chimpanzees. They are tailless, with long arms that touch the ground as they stand semi-erect. The gorillas of Africa may reach a height of five feet and are of great strength; the chimpanzees are slighter but have the highest mental powers and can be trained in a way that betrays a real degree of intelligence.

EDENTATES

Sloths, armadillos, ant-eaters, aard-varks and pangolins form a weird assembly of unfamiliar animals, apparently different from one another but seen to be related through their common ancestry. All, however, have a peculiar dentition. Ant-eaters have no teeth: the rest have cheek teeth all alike and of a simple type with very little enamel. This is scarcely a primitive feature, but a more primitive characteristic is the permanent retention of the testes in the abdomen. Features that are reminiscent of reptiles are the external bony shield of the armadillo and the overlapping scales of the pangolin.

LAGOMORPHA

Rabbits and hares were formerly included with other gnawing animals in the order Rodentia; their most easily detected differences are their long ears and short tails and in addition they have a second pair of small incisor teeth behind those of the upper jaw. The surface of the squamosal, with which the mandible articulates, is broad and flat, imparting considerable freedom of movement to the lower jaw, a freedom that can easily be seen by anyone who watches a rabbit feeding. Another and equally characteristic feature is the great strength of the hind legs, which makes the running of hares and rabbits into a series of leaps.

Hares, of the genus Lepus, include between twenty and thirty species on the continent of Europe. There are also hares native to India and South Africa, while in America the cottontails form the genus Sylvilagus. Rabbits in the genus Oryctolagus, do not show so many species. Originating in North Africa and South Europe they are now widespread.

RODENTIA

The true rodents, with a single pair of upper incisors, include squirrels, rats, voles and mice, porcupines and beavers with so many others that they are the largest and most widely distributed order of mammals. Further characteristics of the teeth continue the distinction from the rabbits. The incisors, with enamel on the front surface only, become chisel-shaped: they have no roots but arise from permanent pulp so that they continue to grow throughout the life of the animal. When the molars bite, those of the upper jaw are inside those of the lower jaw, whereas with Lagomorpha they are reversed. The two sides of the lower jaw are not fused together and by muscular action can be slightly separated, so that the incisor teeth may act like pincers, a very useful function. The limbs are plantigrade and the fore-paws often hold things as in a hand.

A great feature in the lives of all rodents is the rapidity with which they can breed, and the almost over-intense sexuality of the males.

As is obvious from the names of rodents given above, the order has representatives in almost every type of habitat. There are mice underground, dormice in trees and squirrels upon them, voles and beavers in water. Squirrels of the family Sciuridae are a large family, related to the flying squirrels of the family Petauristidae. These are more accurately gliders, which extend their leaps by means of a fold of skin on each side between the fore- and hind-legs. Rats and mice constitute the largest mammalian family, the Muridae, universally distributed. Jerboas, porcupines, and many other species vary in size from the great rat Cricetomys some two-and-a-half feet long, to tiny mice; and include the popular hamster Cricetus cricetus.

CETACEA

Whales, dolphins and porpoises, the only completely marine order of mammals, have always attracted their full share of interest, partly because of the great size of the whales and partly because of the great value to man of much that the whale's body can supply.

Their mode of life is probably responsible for the parallel evolution that has given them a streamlined shape, essentially like that of a fish, while at the same time they have lost several typically mammalian features. They have no hair, save for a few scattered bristles, and the insulation which hair gives to mammals on land is supplied by a thick layer of subcutaneous fat or blubber. The ears have no external pinnae, the eyes no nictitating membrane, the skin has neither muscles nor sweat glands, the testes have no scrotum, the mouth no saliva. There are no hind-legs and the pelvic girdle is represented by a vestigial ischium. The pectoral girdle has no clavicle and the fore-limbs are paddle-like flippers. The tail is converted into a fin of two flukes set horizontally.

The largest animal, living or extinct, that the world has seen is the blue whale, Balaenoptera sibbaldi (or B. musculus) which reaches a length of ninety or even a hundred feet.

The group, which is probably evolved from a carnivorous ancestry, is very sharply distinct from every other mammalian order.

There are two sub-orders:

Odontoceti: toothed whales
Mystaceti (Mystacoceti):   whalebone whales

The toothed whales include the sperm whale, Physeter macrocephalus, the killer whale or grampus, Orca gladiator, as well as the dolphins and propoises. They have teeth in one or both jaws, sometimes as many as 120 in each jaw, as in the common dolphin. Their food consists of fish and other prey often of considerable size.

The whalebone whales, on the other hand, have no teeth, save for a short time before birth. Whalebone or baleen grows in triangular plates, perhaps more than two hundred on each side of the mouth, and forms a remarkable and efficient filter. This allows the escape of water taken in through the open mouth, while retaining all small animals that have entered at the same time. These are jerked off the plates on to the tongue and swallowed. The right whales, humpbacks and rorquals represent the chief families.

SIRENIA

The dugong and manatees are the only survivors of this unusual order, often known as sea cows. Both are aquatic but have little relationship to the whales, their resemblance to which is due only to convergence. Their nearest relatives are probably Hyrax, the coney and the elephants.

They are slow-moving animals with heavy bodies and massive skeletons, some hair is present on their skins, but like whales, they have no hind-legs and no external ears. They are harmless herbivores that feed on seaweed and aquatic plants, a habit which has endowed them with a complex stomach, long intestine, a caecum, and an absence of canine teeth.

There are several species of dugong belonging to the genus Halicore found off the coasts of Africa, India and Australia; the manatees of the genus Manatus occur in American as well as African waters. They never swim far from the shore, but they have been known to ascend rivers, and manatees have been reported far up the Amazon, and in Africa as far as Lake Chad.

They are an interesting group of peculiar mammals, with the unusual distinction of being responsible for the myth of the mermaid.

CARNIVORA

Although this order has but six or seven hundred species many of its members are much more widely known than are those of the larger order of rodents. It includes the lions, tigers and other cats, the foxes, wolves and other dogs, the badgers and the bears.

As their name implies, they are almost without exception eaters of flesh, and most of them are bold predators who leap upon their victims and devour them immediately. Their speed is often considerable, their senses acute, and their intelligence remarkable.

All their feet are provided with claws, which are sometimes retractable, but their main weapons are their long, sharply-pointed canine teeth. The molars have pointed cusps, and a characteristic feature is found in the last upper premolar and the first lower molar, which are large teeth meeting each other with the shearing action of the blades of a pair of scissors. Correlated with this is the closeness of the connection between the dentary or lower jawbone and glenoid, which prevents any of the lateral movements of the jaw to be seen in the Lagomorpha. In a badger's skull the lower jaw remains attached to the rest of the skull even after the whole has been cleaned and bleached.

It may be said of the soft tissues that the stomach is simple and the caecum is often absent, in accordance with the largely protein diet.

PINNIPEDIA

Seals and walruses were formerly placed in a sub-order of the Carnivora, contrasting with the order just described. Their outstanding difference is their adaptation to a marine life, yet apart from this their resemblance to the terrestrial carnivores is so close that Lamarck was unable to say whether the dog is a seal adapted to the land or a seal is a dog adapted to the water. One very obvious sign of their relationship is the intelligence of seals and their response to education and training, in which they are quite the equal of dogs.

They remain flesh-eaters, accustomed to feed on fish, molluscs and crustaceans. They have not the carnassial arrangement of the true carnivores, and the cusps of their molars often point backwards. Unlike the Cetacea they have retained their hind-legs. The humerus and femur are inside the skin of the body, but both feet and hands have all five digits. The tail is short, the clavicle is absent.

In the Otariidae, the most primitive family, the body is covered with fur and there is a small external ear: in these and other respects a relationship to the Fissipedia is shown. The most specialized family, the Phocidae, have no ears and retain their testes in the abdomen. The hind-legs project backwards and the animal progresses on land by a series of jumps. Between these families, the Trichechidae or walruses have long tusks, their upper canines, and can turn their feet forwards when on land. They reach a great size, and lead gregarious lives.

PROBOSCOIDEA

Elephants occupy an order of their own which, among living animals, is represented by three species only, the Indian, African and Chinese elephants.

The general appearance and habits of at least the first two of these are well-known to most readers. Their zoological interest lies chiefly in the specializations of their skull, proboscis and teeth.

The large size of the skull enables it to support trunk and tusks. Many of the bones contain air-spaces in connection with the nasal passage, and the brain is well developed.

The trunk is a unique development of the nose, with the nostrils at its tip. The tusks are the two upper incisors, of almost pure ivory, with enamel at the tip only, where it very soon wears away. There are no canine teeth and no premolars. Each side of each jaw can produce six molars, but as a rule only one of these is functional at any one time.

The stomach is relatively simple for a herbivore, but the caecum is large. The testes remain in the abdomen, except during the mating season.

The two more familiar species are distinguished chiefly by the larger size of the ears in Loxodonta africanus: the great area facilitates the loss of heat from the huge body.

ARTIODACTYLA

The story of mammalian specialization reaches its climax in the two orders of hoofed animals, formerly put together in an order called Ungulata, but now separated into two orders, the Artiodactyla and the Perissodactyla. The obvious difference between them is the number of remaining toes: the Artiodactyla are the even-toed ungulates which have kept the third and fourth digits of each foot, the Perissodactyla are the odd-toed ungulates with the third digit the largest and only functional toe.

There are other features. In the Artiodactyla, which include sheep, goats and deer as well as domestic cattle, pigs and camels, there is very little difference between premolar and molar teeth, and the thoracic and lumbar vertebrae number nineteen. The complex stomach is a great characteristic, and may consist of rumen or paunch, reticulum or honeycomb, psalterium or manyplies, and abomasum or reed. The common names for these parts bear witness to their popularity with eaters of tripe; in the animal possessing them they represent the ability to swallow a quantity of food, regurgitate it in small amounts while resting at ease, and masticate it afresh. Chewing the cud is so familiar and is so easily observed in the farm that it is liable to lose the emphasis that by rights should fall on a remarkable habit, evolved by no other order of mammals.

The graminivorous habit of the order is accompanied by a tendency to produce outgrowths from the frontal bone, which, covered with keratin, constitute the horns and antlers so distinctive of the order.

To these characters there may be added a relative limitation of leg movement to a longitudinal direction, an adaptation that leads to speed, and the loss of the clavicle, so common among the highest orders.

PERISSODACTYLA

It would be pointless to try to decide whether this order or the last is to be regarded as the most specialized or most highly evolved. They represent parallel paths of evolution in which several features in one group are found in the form of their alternatives in the other, and both sets of combinations are efficient animals.

The odd-toed ungulates (horses, rhinoceroses, tapirs) walk,

run or gallop on the middle or third digit, sometimes accompanied by vestiges of the second and fourth, and their speed is possibly due to the raising of the foot so that the metatarsals and metacarpals are almost vertical. The diet is vegetarian, but the cheek teeth are divisible into premolars and molars, while canine teeth ('tushes') are present in many of the males. There are twenty-three thoracic and lumbar vertebrae.

The stomach is much simpler than in those ungulates that chew the cud, and digestion is assisted by the bacteria from the large caecum. There are never any horns or outgrowths from the frontal bone.

In all these respects the Perissodactyla differ from the Artiodactyla. The order does not show the same range of form, and only three families are recognized. These are the Tapiridae, the Equidae and the Rhinocerotidae.

INDEX

235